Oxford International Primary

Maths
Practice Book

Tony Cotton

Language consultants:
John McMahon
Liz McMahon

OXFORD

Contents

1 Numbers and counting

What students will learn

This unit introduces numbers, the number system and counting. The focus is on numbers 0–20. This unit also teaches students how to make estimates. Estimating is an important skill to help with mental calculation and is important in our everyday lives.

Learning objectives:

*   count, read and write numbers to 100
*   count in multiples of twos, fives and tens
*   know and make numbers using objects and pictures
*   use words such as equal to, more than, less than (fewer), most, least
*   read and write numbers from 1 to 20 in words.

Key words

coin	count on	10 more	smallest
numbers	altogether	10 less	largest
number names	how many?	ones	multiple
zero–twenty	most	twos	steps
number	fewest	fives	predict
count	1 more	tens	estimate
more than	1 less	cubes	guess
less than	2 more	rods	nearly
fewer than	2 less	order	how many more?
count up to	5 more	column	how many less?
count back	5 less	row	

Ways to help

*   Ask students to count lots of different objects. Always start from 1 so that students learn to say the number names in order, e.g. *How many cars on the street? How many cups on the table? How many items in the shopping basket?*
*   Emphasise the last number you count so students learn that the last number tells you how many objects there are, e.g. *There are one, two, **three** cups.* Touch objects as you count them so students understand you only count each object once.
*   Once students can count on it is helpful to count back too. This helps in understanding the pattern and is useful later when beginning to subtract.
*   When students count groups of objects, ask them to think about which group has the most and the fewest objects. Encourage them to compare, e.g. *This group has more, this group has fewer.*
*   Say counting rhymes together.
*   If students find it hard to count pictures, give them a matching number of objects to count instead.
*   Keep handy images of number lines and 100-squares so that students have a range of mental images of the number system. This is particularly helpful in mental calculation as they can visualise these images in their heads.

2 Number bonds

What students will learn

This unit continues to teach students about the number system. Students will learn how to use their knowledge of the number system to solve problems. This unit introduces students to addition and subtraction by combining and separating sets of objects into number bonds.

Learning objectives:

- read and write number sentences using addition (+), subtraction (–) and equals (=) signs
- use number bonds to 20 and related subtraction facts.

Key words

add (+)	total	altogether	count on
addition	how many?	number	count back
subtract (–)	how many more?	missing number	less
subtraction	how many less?	number pairs	more
equals (=)	partition	number bonds	

Ways to help

- Continue to count aloud with students. This helps them understand that number names are always said in the same order and that the number of objects in a group is given by the last number that we say when counting.
- Use practical materials when counting, such as counters, cubes or buttons. Use the objects to add two groups together and to split the whole set into two groups. This is the best way for students to begin to see what we mean by addition (how many altogether) and subtraction (how many left).

3 Exploring numbers

What students will learn

This unit introduces ordering numbers and finding numbers that are more than a given number, less than a given number and between two given numbers. Place value is explored through partitioning 2-digit numbers into tens and ones, and discussing the value of given digits.

This unit also includes some work on ordinal numbers, and explores the patterns of even and odd numbers.

Learning objectives:

*   know and make numbers using objects and pictures
*   read and write numbers from I to 20 in words
*   say I more and I less than any number
*   count to 100 and beyond.

Key words

more than	ones	second	odd
less	twos	third	even
fewer than	fives	fourth	predict
how many?	tens	fifth	column
between	place-value table	after	row
highest	2-digit number	before	
lowest	ordinal numbers	pattern	
partition	first	repeating pattern	

Ways to help

*   It is important to use practical equipment to help students count and compare numbers at this stage, including objects that can be grouped, such as straws or buttons. It is particularly important to give students opportunities to group objects in tens.
*   Continue to count objects aloud, asking students to touch each object as it is counted.
*   Create a number line from 0–20 and put it on a wall. Encourage students to find numbers on the line and then to find larger and smaller numbers.

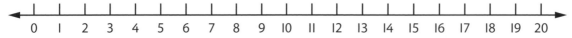

*   Encourage students to look for number patterns in the environment, such as odd and even house numbers on either side of the road.

4 Addition

What students will learn

This unit introduces students to the idea that addition is what happens when the objects in two or more smaller groups come together to make one big group, and that changing the order of addition does not change the total.

Students are taught various strategies for solving addition problems, including the use of number lines and known facts such as number bonds. Students apply their learning to solve word problems involving addition.

Learning objectives:

· read and write number sentences using addition (+), subtraction (–) and equals (=) signs
· use number bonds to 20 and related subtraction facts
· add I-digit and 2-digit numbers
· solve addition problems.

Key words

addition	more	number sentence	equals (=)
add (+)	pairs	sum	jump
altogether	total	plus	bridging I0

Ways to help

· Use practical materials to help students see the commutative property of addition. Commutative means that it does not matter which way round we add two numbers. For example, students might be given a word problem such as: *I have 3 toy cars and my friend has 8 toy cars. How many do we have altogether?* This can be solved in two ways:

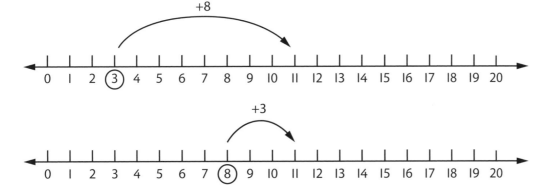

· Students can use 0–20 number lines when solving addition problems. For example, to solve $3 + 8$, or $8 + 3$:

5 Subtraction and difference

What students will learn

This unit introduces students to the idea of subtraction and difference.

Students will be taught about partitioning to help them make the links between addition and subtraction. For example: *I can partition 7 into 4 and 3, so 14 – 7 is the same as 14 – 4 – 3 = 10 – 3.* This helps students see that 14 – 7 = 7.

They will also learn that they can check the answer to a subtraction by adding. For example: *I know that 8 – 5 = 3 is correct because 3 + 5 = 8.*

Students are taught various strategies for solving subtraction and difference problems, including the use of number lines and known facts such as number bonds. Students apply their learning to solve word problems involving subtraction.

Learning objectives:

- read and write number sentences using addition (+), subtraction (–) and equals (=) signs
- subtract 1-digit and 2-digit numbers
- solve subtraction problems.

Key words

difference	count back	take away	how many less?
minus	count on	how many left?	number bonds
subtract (–)	jump on	how many more?	
subtraction	jump back		

Ways to help

- Help students try to spot patterns and notice the relationship between addition and subtraction. For example:

  8 + 5 = 13
  13 – 8 = 5
  13 – 5 = 8
- Use practical materials to show students these patterns.

6 Multiplication and division

What students will learn

This unit introduces students to multiplication and division. Division is introduced as grouping and sharing, and multiplication is illustrated using repeated addition and arrays. An array has an equal number of rows and columns. For example, this array shows $2 \times 3 = 6$ and $3 \times 2 = 6$:

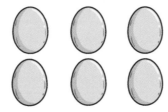

Students apply their learning to solve word problems involving multiplication.

Learning objectives:

*   solve multiplication and division problems
*   use pictures and arrays to solve multiplication and division calculations.

Key words

share	column	lots of	multiples of 2
equally	even	multiply ($\times$)	multiples of 4
groups	odd	multiplication	multiples of 5
equal groups	halve	divide ($\div$)	repeated addition
array	half	division	
row	groups of	multiple	

Ways to help

*   While playing games, encourage students to see multiplication number patterns on game equipment such as dominoes. For example, 6 dots could be 3 groups of 2 or 2 groups of 3.
*   Continue to use the 100-square (see the example on page 27) to help students to see patterns when counting.
*   Encourage students to look for arrays in everyday life. For example, this egg box shows $6 \times 2 = 12$ and $2 \times 6 = 12$. It also shows $12 \div 2 = 6$ and $12 \div 6 = 2$.

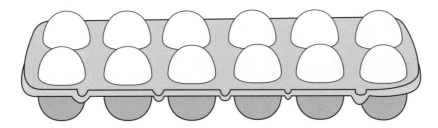

7 Fractions

What students will learn

This unit introduces students to fractions. They will learn about halves ($\frac{1}{2}$) and quarters ($\frac{1}{4}$). The most important thing for students to understand is that when we divide shapes or amounts into fractions, all the parts must be of equal size.

Learning objectives:

- find a half of an object, shape or quantity
- find a quarter of an object, shape or quantity.

Key words

double	quarter ($\frac{1}{4}$)	equal parts
halve	three quarters ($\frac{3}{4}$)	equal pieces
half ($\frac{1}{2}$)	one out of four	equal groups
halves	fraction	

Ways to help

- Encourage students to look for 1-digit numbers in the local environment, and ask them to double or halve the numbers.
- Practical experience is always helpful. Give students opportunities to practise cutting shapes into halves and quarters and ask them to check that the parts are the same size. Similarly, ask students to divide a quantity of objects (such as pieces of fruit, pasta or counters) into halves and quarters by sharing them into equal groups.
- At home, parents and carers can use the vocabulary of fractions when they are sharing things between family members, e.g. *You and your sister can have half each.*

8 Length, mass and capacity

What students will learn

This unit introduces students to early ideas of measurement. They will learn about lengths and masses of objects by comparing them. They will estimate the capacities of containers (how much they hold) and then check their estimates and comparisons using small objects such as dried chickpeas. They will also be introduced to comparative language, such as longer, shorter, heavier, lighter, and so on. They will also start to use measuring equipment.

Learning objectives:

- solve length, height, mass and capacity problems
- measure and begin to record length, height, mass and capacity.

Key words

length	mass	holds the most
width	large, larger, largest	holds the least
long, longer, longest	small, smaller, smallest	how many?
short, shorter, shortest	heavy, heavier, heaviest	compare
wide, wider, widest	light, lighter, lightest	measure
tall, taller, tallest	capacity	measuring jug

Ways to help

- Students will spend a lot of time in class thinking about the conservation of measures. That is the idea that an object keeps the same size and shape no matter how you position it. Give students plenty of practice at comparing the lengths of objects, making sure they line up the objects so that the ends are together.

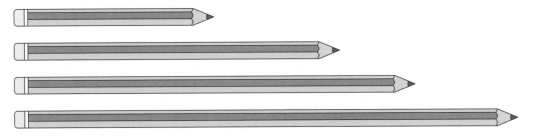

- Carry out lots of practical measuring activities, e.g. measuring and comparing lengths and asking which is longer and which is shorter.
- Ask students to use scales to weigh things and decide which is heavier and which is lighter.
- Ask students to measure out volumes of liquids using measuring jugs.

9 Money

What students will learn

This unit introduces students to the values of notes and coins. They will learn about the currency of their own country as well as US dollars and cents and UK pounds and pence. They will explore how to make given amounts using different combinations of coins.

Learning objective:

- recognise different coins and notes.

Key words

coin	cents	price	greater than
note	pounds	total	stamps
currency	pence	how much?	postage
dollars	cost	less than	value

Ways to help

- Show students the coins that are used in your currency. Help students to understand the value of the different coins.
- Talk about how much things cost when going shopping. Look at price labels and till receipts.
- Play games to find different ways to make small amounts of money with different coins.

10 Time

What students will learn

Telling the time is a vital skill for young students to learn. It is complicated, as time is not organised according to the decimal system. Instead of counting in tens, there are 60 seconds in a minute, 60 minutes in an hour and 24 hours in a day.

Students start telling the time by recognising and naming different parts of the day, such as morning, afternoon and night. In this unit, they learn the names of the days of the week, and begin to measure time in minutes and seconds.

Learning objectives:

- tell the time to the hour and half past the hour
- use the words before, after, next, first, today, yesterday, tomorrow, morning, afternoon and evening
- know the days of the week and the months of the year
- measure time (hours, minutes, seconds)
- solve time problems.

Key words

days of the week	tomorrow	hour hand	next
day	o'clock	minute hand	morning
week	hours	what time is it?	afternoon
month	minutes	how long?	evening
calendar	seconds	before	night
yesterday	short hand	after	
today	long hand	first	

Ways to help

- It is useful to have both digital and analogue clocks in the classroom and at home. Look out for when the time is either an o'clock time or half past the hour and ask students to tell you what time it is.
- Use the vocabulary of time when talking to students about plans for the day, or daily routines, e.g. *What did we do yesterday? What would you like to do tomorrow?*

11 Geometry

What students will learn

This unit will help students develop the vocabulary to describe the properties of the shapes that they see every day. They can then use these properties to start to classify (sort out) shapes. They will start noticing whether shapes are regular (have all the sides and angles the same) or irregular (do not have all the sides and angles the same).

Learning objectives:

- recognise and name 2D shapes
- recognise and name 3D shapes
- describe position, direction and movement, including turns
- identify line symmetry.

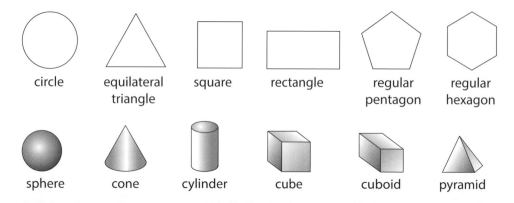

circle   equilateral triangle   square   rectangle   regular pentagon   regular hexagon

sphere   cone   cylinder   cube   cuboid   pyramid

Key words

vertex (corner)	cuboid	line of symmetry	between
edge	cylinder	symmetrical	above
side	sphere	reflection	on top of
face	straight edge	mirror line	below
square	curved edge	front	under
rectangle	flat	back	turn
triangle	2-dimensional (2D)	in front of	whole turn
circle	3-dimensional (3D)	behind	half turn
cube	line symmetry	next to	quarter turn

Ways to help

It is important that students spend a lot of time on practical activities. For example, they could:

- notice and play with everyday shapes
- sort shapes in different ways
- most importantly, talk about what they notice about the shapes.

12 Statistics

What students will learn

There are different ways of sorting, organising and representing information (data) that help us make sense of information. As students work through the *Oxford International Primary Maths* series, they will learn about 'the data handling cycle', shown here.

Start with a question
What do we want to know and what information can we collect to help answer the question?

Collect the data
Find out answers to the questions we posed, by asking people.

Sort and represent the data
Create a chart or graph that helps us make sense of the data.

Interpret the results
Answer the original question.

It is called a cycle as the final step usually results in us asking more questions.

This unit introduces students to some simple forms of representing data: tables, pictograms, block diagrams, Venn diagrams and Carroll diagrams.

Learning objectives:
- use block graphs and pictograms
- sort objects using Venn and Carroll diagrams.

Key words

data	column	Venn diagram	sort
list	chart	Carroll diagram	criteria
table	pictogram	most	intersection
row	block diagram	fewest	

Ways to help

- Look out for examples of charts and graphs in magazines, on the Internet and in the local environment. Talk about these with students.
- You can also ask simple questions about preferences, such as the types of foods different people like best, or favourite places to visit.

1A Counting objects

Discover   Student Book 1, page 7

- colouring pencils

Colour squares to match the number.

Count the coloured squares out loud.

Write the numbers under the squares.

The first one is done for you.

You could count cubes (or other small objects) before you colour the squares.

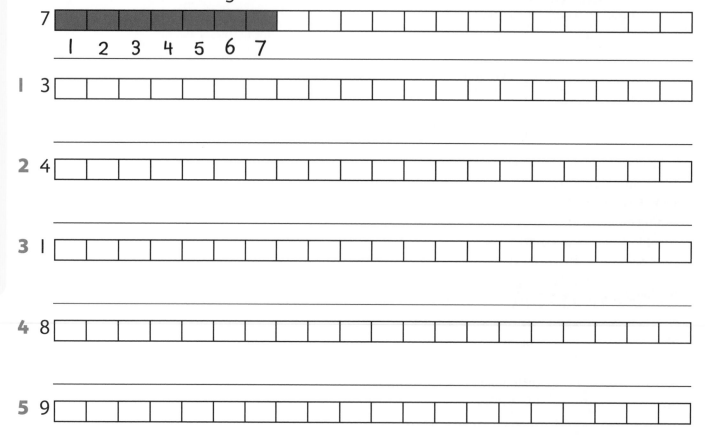

7 ▣▣▣▣▣▣▣ ☐☐☐☐☐☐☐☐☐☐☐☐☐☐☐☐☐

     1   2   3   4   5   6   7

1   3

2   4

3   1

4   8

5   9

Stretch zone

How many more coloured squares are there in the first row than in the second row? ☐

1A Counting objects

Explore 1   Student Book 1, page 8

- Look at the number at the top of each box.
- Count out loud the correct number of counters (or other small objects).
- Draw the correct number of counters in the box.

The first one is done for you.

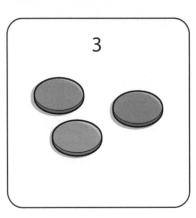

3

1
| 7 |

2
| 13 |

3
| 14 |

4
| 10 |

5
| 16 |

6
| 8 |

 Stretch zone

Draw a ✔ next to the box with the **most** counters.

Draw a ✘ next to the box with the **fewest** counters.

How many more counters are in the box with the most than in the box with the fewest?

Explore 2   Student Book 1, page 9

- colouring pencils: blue and red

Draw the number of beads on each string. The first one is done for you.

9

1 12

2 7

3 18

4 13

5 4

6 15

7 11

8 Draw a ✔ next to the string with the most beads.

Stretch zone

Make a pattern. Draw beads on the string and colour them red and blue.

There must be 3 more red beads than blue beads.

Discover 1 Student Book 1, page 10

Draw pictures to match the numbers. Write the numbers in words.

The first one is done for you.

	4 cars		four
1	11 worms		
2	13 squares	□	
3	17 triangles	△	
4	9 people		
5	8 circles	○	
6	20 faces	☺	

Stretch zone

There are seven cats.

One cat runs away.

How many cats are left?

Draw the cats that are left.

Discover 2   Student Book 1, page 11

Draw rods and cubes to show these numbers. The first one is done for you.

	23	⬛⬛⬛⬛⬛⬛⬛⬛⬛⬛ ⬛⬛⬛⬛⬛⬛⬛⬛⬛⬛ ☐ ☐ ☐
1	31	
2	36	
3	26	
4	25	
5	43	
6	47	

Stretch zone

Write two different numbers and draw the rods and cubes.

Explore 1   Student Book 1, page 12

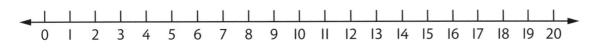

0 1 2 3 4 5 6 7 8 9 10 11 12 13 14 15 16 17 18 19 20

Use the number line to answer these questions.

The first one is done for you.

I start at 3. I add 2.   `5`

1 I start at 5. I add 7.

2 I start at 5. I add 6.

3 I start at 5. I add 5.

4 I start at 11. I add 3.

5 I start at 1. I add 3.

6 I start at 14. I add 4.

7 I start at 13. I add 4.

8 I start at 12. I add 4.

Stretch zone

Write three different addition sentences with the answer 19.

_____   _____

_____

Explore 2   Student Book 1, page 13

Write the missing numbers on the lily pads. The first one is done for you.

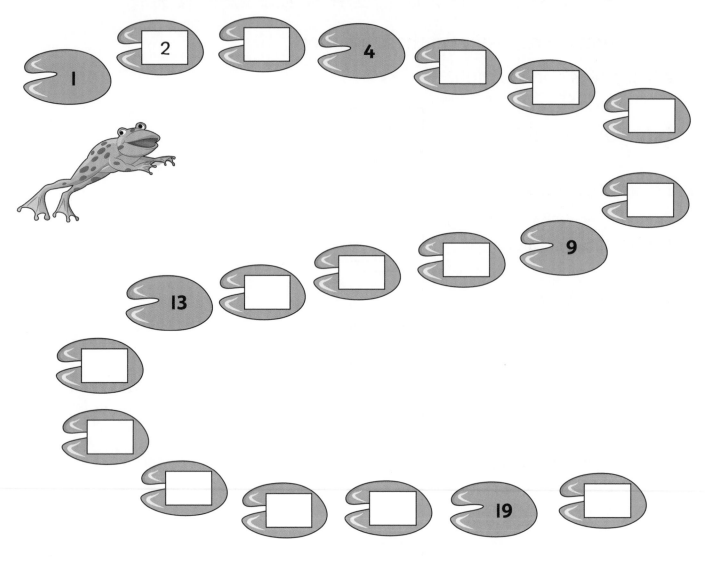

🎯 **Stretch zone**

Write the words for the seven numbers that end with 'teen'.

_____

_____

_____

1B Reading and writing numbers

Explore 3   Student Book 1, page 14

Follow the instructions to make numbers.

Draw rods and cubes to show each number.

The first one is done for you.

> You could use a number line to find the numbers. There are lots of possible answers each time.

		Number	Rods and cubes
	larger than 20	23	
1	smaller than 25		
2	larger than 14		
3	smaller than 18		
4	larger than 32		
5	smaller than 31		

Stretch zone

Write a number between 35 and 40. Draw it with rods and cubes.

Number	Rods and cubes

Discover 1   Student Book 1, page 15

Draw the jumps on the number lines. Write the numbers you land on.

The first one is done for you.

I start at 3. I jump on in twos. I land on <u>5, 7, 9, 11, 13, 15, 17, 19</u> .

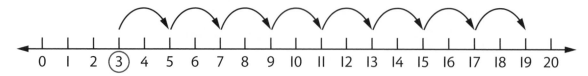

I   I start at 4. I jump on in twos. I land on _____ .

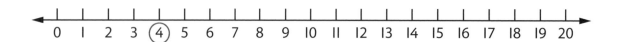

2   I start at 19. I jump back in twos. I land on _____ .

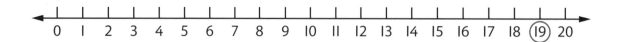

3   I start at 16. I jump back in twos. I land on _____ .

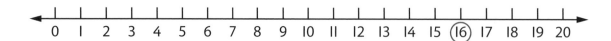

⊙ **Stretch zone** ▶

Draw your own jumps on or back on this number line. Write a sentence to describe the jumps.

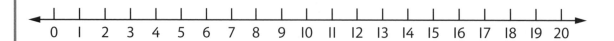

_____

Discover 2   Student Book 1, page 16

Draw the jumps on the number lines. Write the numbers you land on.

1 I start at 2. I jump on in tens. I land on _____.

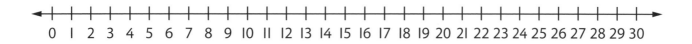

2 I start at 29. I jump back in tens. I land on _____.

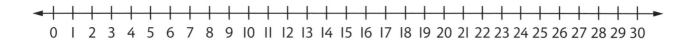

3 I start at 8. I jump on in tens. I land on _____.

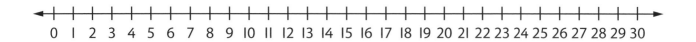

4 I start at 23. I jump back in tens. I land on _____.

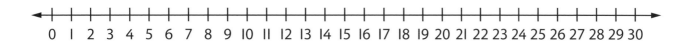

5 I start at 7. I jump on in tens. I land on _____.

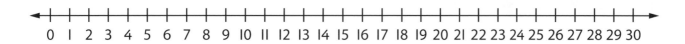

Stretch zone

Use your number lines to help you answer these calculations.

$8 + 10 =$ ☐ $23 - 10 =$ ☐

$18 + 10 =$ ☐ $13 + 10 =$ ☐

1 Numbers and counting

Discover 3   Student Book 1, page 17

Draw the jumps on the number lines. Write the numbers you land on.

1   I start at 2. I jump on in fives. I land on _____ .

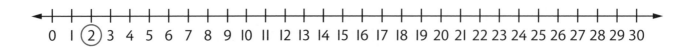

2   I start at 4. I jump on in fives. I land on _____ .

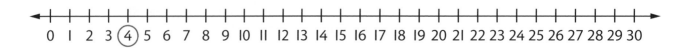

3   I start at 26. I jump back in fives. I land on _____ .

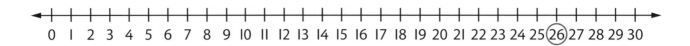

4   I start at 27. I jump back in fives. I land on _____ .

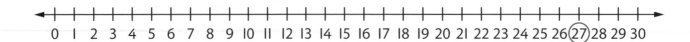

Stretch zone

Draw your own jumps on or back on this number line. Write a sentence to describe the jumps.

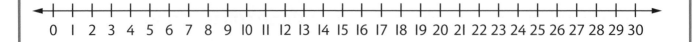

_____

_____

Explore 1   Student Book 1, page 18

Draw the jumps on the number lines. Write the numbers you land on.

1 I start at 1. I jump on in twos. I land on _____ .

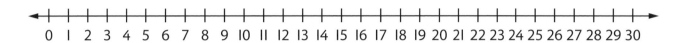

2 I start at 3. I jump on in tens. I land on _____ .

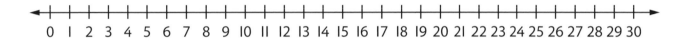

3 I start at 4. I jump on in twos. I land on _____ .

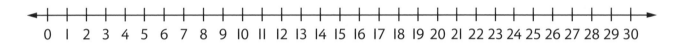

4 I start at 4. I jump on in tens. I land on _____ .

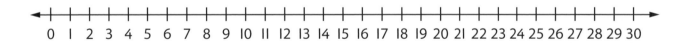

5 I start at 5. I jump on in tens. I land on _____ .

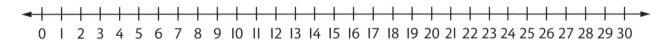

Stretch zone

I start at 1 and jump on in fives. What numbers will I land on?

_____

How do you know?

_____

Explore 2   Student Book 1, page 19

Write the numbers that are 2 less and 2 more than each number.

You could use a number line to help you.

The first one is done for you.

	2 less	Number	2 more
	15	17	19
1		13	
2		20	

Write the numbers that are 5 less and 5 more than each number.

	5 less	Number	5 more
3		25	
4		24	
5		14	

Write the numbers that are 10 less and 10 more than each number.

	10 less	Number	10 more
6		25	
7		24	
8		14	

◎ **Stretch zone**

What patterns do you notice in your answers to 10 more and 10 less?

_____

_____

Explore 3   Student Book 1, page 20

- colouring pencils: red, yellow, blue, green

1	2	3	4	5	6	7	8	9	10
11	12	13	14	15	16	17	18	19	20
21	22	23	24	25	26	27	28	29	30
31	32	33	34	35	36	37	38	39	40
41	42	43	44	45	46	47	48	49	50

Follow the instructions to draw coloured dots next to numbers on the grid.

1 Start at 25 and count on in twos. Draw a red dot in each of these squares.

2 Start at 38 and count back in twos. Draw a yellow dot in each of these squares.

3 Start at 5 and count on in fives. Draw a blue dot in each of these squares.

4 Start at 50 and count back in tens. Draw a green dot in each of these squares.

Stretch zone

Describe the patterns you notice.

_____

_____

Explore 4   Student Book 1, page 21

Write the numbers that are 10 less and 10 more than each number.

Use the 100-square to help you.

The first one is done for you.

	10 less	Number	10 more
	14	24	34
1		47	
2		19	
3		42	
4		58	
5		88	
6		90	
7		45	
8		71	
9		28	
10		36	

1	2	3	4	5	6	7	8	9	10
11	12	13	14	15	16	17	18	19	20
21	22	23	24	25	26	27	28	29	30
31	32	33	34	35	36	37	38	39	40
41	42	43	44	45	56	47	48	49	50
51	52	53	54	55	56	57	58	59	60
61	62	63	64	65	66	67	68	69	70
71	72	73	74	75	76	77	78	79	80
81	82	83	84	85	86	87	88	89	90
91	92	93	94	95	96	97	98	99	100

Stretch zone

Explain how you can use a 100-square to find 10 more or less than a number.

_____

_____

Discover Student Book 1, page 22

> ● four different sorts of small objects that you can hold in your hands

How many of each object do you think you can hold?

Write your estimate and then take a handful to find out.

Was your estimate more or less than the actual number? How many more or less?

An example is shown in the table.

Object	Estimate	Actual number	More or less
cherries	11	8	My estimate was 3 more.

Stretch zone ➤

Did your estimates get better each time? _____

If they did get better, can you explain why?

_____

Explore 1   Student Book 1, page 23

- four small different-sized containers
- some large beans or dried pulses, such as chickpeas

For each container, estimate how many beans or chickpeas you think it will hold.

Write your estimate.

Fill the container and count the actual number.

An example is shown in the table.

Container	Estimate	Actual number	More or less
container	50	37	My estimate was 13 less.
container 1			
container 2			
container 3			
container 4			

Stretch zone

Find another container. Use your answers to estimate how much it holds. Check your estimate.

Estimate: [ ]   Actual number: [ ]

More or less? _____

1D Estimating

Explore 2   Student Book 1, page 24

Estimate the number of dots. Do not count them!

Draw a circle around the number that you think is a good estimate.

1 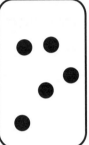   2   5   9

4    6   9   12

2 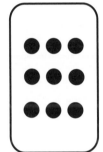   9   12   14

5 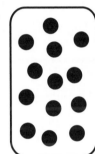   10   13   16

3    3   4   6

6 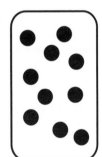   5   8   10

 Stretch zone

How many sweets do you think there are in this jar? ☐

How did you make your estimate?

_____

_____

Review

1 Draw a face next to each bubble to show how you feel about your learning.

counting objects

reading and writing numbers

counting in twos, fives and tens

estimating quantities

2 Tell a partner about one thing you did really well in this unit.

3 Draw or write about things you found easy, challenging or really hard.

What work did you feel confident doing?

What work was challenging?

Is there any work you might need some extra help with?

2A Number bonds for 6, 7, 8, 9

Discover   Student Book 1, page 28

- colouring pencils: two different colours

You could use cubes (or other small objects) to help you.

Colour circles to match the number pairs.

The first one is done for you.

1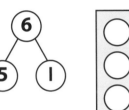

4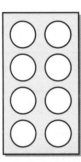

2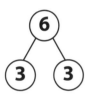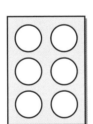

5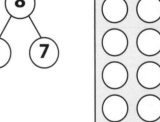

3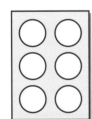

6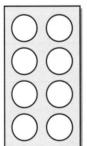

Stretch zone

How many different number pairs can you find to make 8?

_____

Explore Student Book 1, page 29

Write all the number bonds for 6, 7 and 8.

	Number bonds for 6				
1	0	+		=	6
2		+		=	6
3		+		=	6
4		+		=	6
5		+		=	6
6		+		=	6
7		+	0	=	6

	Number bonds for 8				
16	0	+		=	8
17		+		=	8
18		+		=	8
19		+		=	8
20		+		=	8
21		+		=	8
22		+		=	8
23		+		=	8
24		+	0	=	8

	Number bonds for 7				
8	0	+		=	7
9		+		=	7
10		+		=	7
11		+		=	7
12		+		=	7
13		+		=	7
14		+		=	7
15		+	0	=	7

Stretch zone

Write all the number bonds for 9.

2B Number bonds for 10

Discover Student Book 1, page 30

Write the missing numbers in these bar models. The first one is done for you.

You could use 10 cubes (or other small objects) to help you.

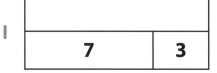

1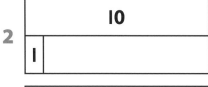

2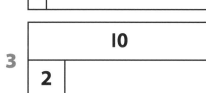

3

4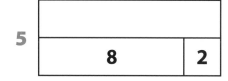

5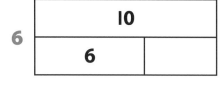

6

7

Stretch zone

Choose four different number pairs for 10. Colour the circles to match the number pair. Write the number sentence underneath.

One is done for you.

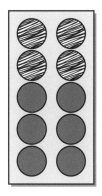 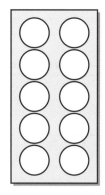 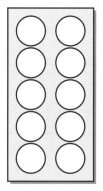 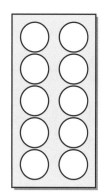 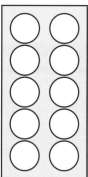

4 + 6 = 10 _____ _____ _____

Explore   Student Book 1, page 31

Draw spots on each of these dominoes to add to 10.

Write the number sentence underneath.

The first two are done for you.

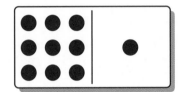

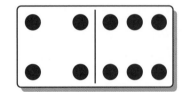

 I

9 + 1 = 10   4 + 6 = 10   _____

2    3    4

5   6   7

Stretch zone

Are there any more dominoes that add to 10?

2C Missing numbers

Discover   Student Book 1, page 32

Write the missing numbers.

Use the 100-square to help you.

The first one is done for you.

1	2	3	4	5	6	7	8	9	10
11	12	13	14	15	16	17	18	19	20
21	22	23	24	25	26	27	28	29	30
31	32	33	34	35	36	37	38	39	40
41	42	43	44	45	56	47	48	49	50
51	52	53	54	55	56	57	58	59	60
61	62	63	64	65	66	67	68	69	70
71	72	73	74	75	76	77	78	79	80
81	82	83	84	85	86	87	88	89	90
91	92	93	94	95	96	97	98	99	100

29 is 2 less than 31.

1 ☐ is 2 more than 13.

2 ☐ is 10 more than 28.

3 ☐ is 1 less than 18.

4 ☐ is 1 more than 18.

5 ☐ is 10 less than 14.

6 ☐ is 2 less than 26.

7 ☐ is 2 more than 20.

8 ☐ is 10 less than 19.

9 ☐ $+ 4 = 10$

10 $3 +$ ☐ $= 9$

11 ☐ $- 7 = 9$

12 $1 +$ ☐ $= 5$

Stretch zone

Write four missing-number sentences. All your number sentences must include the number 15.

_____   _____

_____   _____

Write number sentences that give the correct answers.

For each question, write two numbers and an addition (+) or subtraction (−) sign.

Try to use a mixture of + and − signs.

The first one is done for you.

$$\boxed{7}\ \bigcirc\!\!\!\!- \ \boxed{5}\ = 2$$

1 $\boxed{}\ \bigcirc\ \boxed{}\ = 15$

2 $\boxed{}\ \bigcirc\ \boxed{}\ = 12$

3 $\boxed{}\ \bigcirc\ \boxed{}\ = 7$

4 $\boxed{}\ \bigcirc\ \boxed{}\ = 2$

5 $\boxed{}\ \bigcirc\ \boxed{}\ = 10$

6 $\boxed{}\ \bigcirc\ \boxed{}\ = 20$

7 $\boxed{}\ \bigcirc\ \boxed{}\ = 11$

8 $\boxed{}\ \bigcirc\ \boxed{}\ = 5$

Stretch zone

Write three additions or subtractions with the answer 5.

_____   _____   _____

Explore 2   Student Book 1, page 34

All the machines need 15 balls altogether. Draw a different combination of balls for each machine.

You could use counters (or other small objects) to help you.

The first one is done for you.

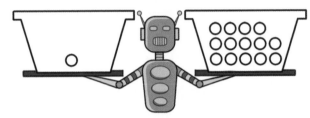

1    2

3 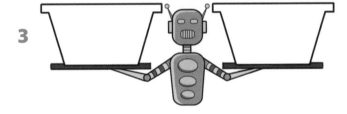   4

Write the missing numbers.

5   7 + ⬚ = 15   8   8 + ⬚ = 15

6   5 + ⬚ = 15   9   3 + ⬚ = 15

7   6 + ⬚ = 15

Stretch zone

Find three numbers that total 15.

⬚ + ⬚ + ⬚ = 15

2 Number bonds

41

Review

1 Draw a face next to each bubble to show how you feel about your learning.

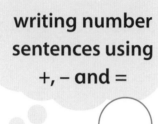

number bonds to 20

writing number sentences using +, − and =

working out missing numbers

2 Tell a partner about one thing you did really well in this unit.

3 Draw or write about things you found easy, challenging or really hard.

What work did you feel confident doing?

What work was challenging?

Is there any work you might need some extra help with?

3A More and less

Discover   Student Book 1, page 38

Complete the sentences by writing how **many** more or **less**.

The first one is done for you.

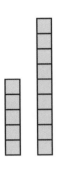

5 is ___4 less___ than 9.

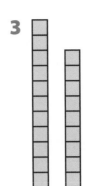

3

11 is _____ than 9.

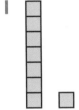

1

7 is _____ than 1.

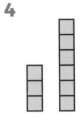

4

3 is _____ than 6.

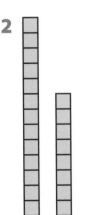

2

13 is _____ than 8.

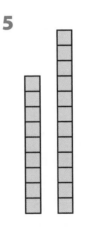

5

9 is _____ than 12.

Stretch zone

Draw a picture to show '4 more than'.

Write a number sentence to describe the picture.

_____

_____

Explore   Student Book 1, page 39

Write the number that is I more
and I less than the given number.

Use the number track to help you.

| I | 2 | 3 | 4 | 5 | 6 | 7 | 8 | 9 | 10 | II | 12 | 13 | 14 | 15 | 16 | 17 | 18 | 19 | 20 |

The first one is done for you.

	I less	Number	I more
	9	10	II
I		II	
2		12	
3		13	
4		14	
5		15	
6		16	
7		17	
8		18	

Stretch zone

Complete these sentences. The first one is done for you.

9 is ___5 more than___ 4.   4 is _____ I.

II is _____ 14.   2 is _____ 9.

17 is _____ II.

3B Between

Discover

Student Book 1, page 40

Write **any** number that is **between** the two numbers.

Use the number track to help you.

1	2	3	4	5	6	7	8	9	10	11	12	13	14	15	16	17	18	19	20

The first one is done for you.

	First number	First number	Second number
	6	7	9
1	8		11
2	12		17
3	15		19
4	9		11
5	1		8
6	16		20
7	5		12
8	11		14

Stretch zone

How many whole numbers are there between 21 and 27? ☐

How many between 31 and 37? ☐

How many between 32 and 38? ☐

Explore   Student Book 1, page 41

Write the missing numbers in these sequences.

The first one is done for you.

16	15	14	13	12	11	10	9	8	7

1

7	8			11		13	14		

2

14	13			10			7		

3

6		8					13		

4

19			16					11	

5

				13	14				

6

								2	1

7

	13				17				

8

		20							13

Stretch zone

Make up your own missing-number puzzle.

3C Tens and ones

Discover Student Book 1, page 42

Partition (split) each number to show the value of the tens and the ones.

Say the number aloud each time.

The first one is done for you.

	Number	Tens value	Ones value
	35	30	5
1	24		
2	52		
3	29		
4	41		
5	19		
6	33		
7	11		
8	46		
9	7		
10	60		

Stretch zone

Write a number that is easy to partition into tens and ones.

Why is it easy to partition?

_____

_____

Explore   Student Book 1, page 43

Write these numbers. The first one is done for you.

	5 tens	3 ones	53
I	2 tens	7 ones	
2	4 tens	4 ones	
3	I ten	8 ones	

Add the tens and ones. Write the number. The first one is done for you.

	40	+	6	=	46
4	20	+	I	=	
5	30	+	5	=	
6	10	+	8	=	
7	20	+	2	=	
8	30	+	6	=	
9	20	+	6	=	
10	10	+	6	=	

Stretch zone

Draw pictures of rods and cubes to show these numbers.

8   18   38

3D Partitioning

Discover   Student Book 1, page 44

For each number, write the value of the tens and the ones.

Then draw rods and cubes for each number.

The first one is done for you.

Number	Tens value	Ones value	Rods and cubes
45	40	5	
28			
18			
17			
7			

1
2
3
4

Stretch zone

Talk to a partner or an adult about any patterns you see in your answers.

3D Partitioning

Explore 1   Student Book 1, page 45

Follow these steps five times.

- Shuffle the digit cards.
- Turn over two cards.
- Use the numbers to complete a row in the table.

- digit cards 1–9

I turned over 4 and 3.

An example is shown in the table.

Cards	2-digit numbers	Number sentences	Which is larger?
4 and 3	43 34	43 = 40 + 3 34 = 30 + 4	43 is larger than 34.

 Stretch zone

What is the largest number you can make with two cards?

What is the smallest number you can make with two cards?

3D Partitioning

Explore 2   Student Book 1, page 46

I pick 5, 6 and 2. I will make 25, 26, 56 and 62.

- 1–6 digit cards (you could make your own)

Choose three cards.

Use the digits to make four different 2-digit numbers.

Write them in the table in order from smallest to largest.

Partition each number into tens and ones.
Draw each number as rods and cubes. An example is shown in the table.

My three cards are ☐, ☐ and ☐.

Number	Partitioned	Rods and cubes
25	25 = 20 + 5	

Stretch zone

Explain how you knew what order to write the numbers in.

_____

Discover   Student Book 1, page 47

Look at the vehicles in this row of traffic. Follow the instructions.

The first one is done for you.

    Draw an arrow below the 10th vehicle.

1   Draw a circle around the 3rd vehicle.   4   Draw a ✘ under the 4th vehicle.

2   Draw a square around the 9th vehicle.   5   Draw a line under the 6th vehicle.

3   Draw a ✔ under the 1st vehicle.

Complete these sentences using ordinal numbers.

6   The caravan   is the _____ vehicle.

7   The motorbike   is the _____ vehicle.

8   The car   is the _____ vehicle.

9   The coach   is the _____ vehicle.

10 The digger   is the _____ vehicle.

Stretch zone

Use toy vehicles to make a traffic jam. Describe it using ordinal numbers.

_____

_____

_____

Explore 1   Student Book 1, page 48

Draw an arrow to match each ordinal number to the correct pattern.

The first one is done for you.

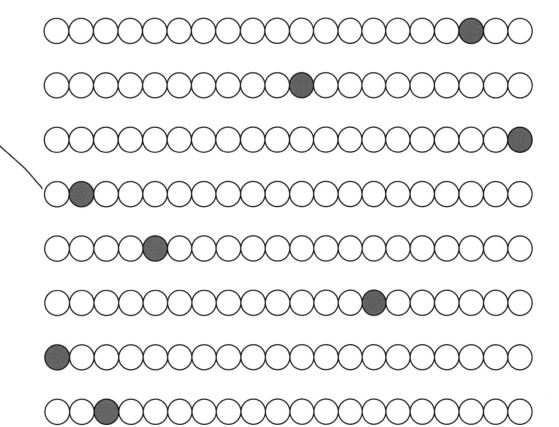

2nd

1 5th

2 11th

3 1st

4 18th

5 9th

6 14th

7 20th

8 3rd

Stretch zone

Colour any three beads. Use a different colour for each bead.

Describe your pattern using ordinal numbers.

_____

_____

3 Exploring numbers

53

Explore 2   Student Book 1, page 49

• colouring pencils

Colour the beads to make your own patterns.

Complete the sentences under each pattern.

The 3rd bead is _____ .

The 9th bead is _____ .

The 12th bead is _____ .

The 18th bead is _____ .

The 4th bead is _____ .

The 7th bead is _____ .

The 17th bead is _____ .

The 14th bead is _____ .

Stretch zone

Colour these beads to make a pattern.

Write two sentences about your pattern using ordinal numbers.

_____

_____

3F Even and odd

Discover   Student Book 1, page 50

• colouring pencils: two different colours

In this number track, start at 2 and use one colour to colour all the even numbers.

Use another colour to colour all the odd numbers.

1	2	3	4	5	6	7	8	9	10	11	12	13	14	15	16	17	18	19	20	21	22	23	24	25	26	27	28

1   What colour are the even numbers? _____

2   What colour are the odd numbers? _____

3   Write eight even numbers in the first circle.

4   Write eight odd numbers in the second circle.

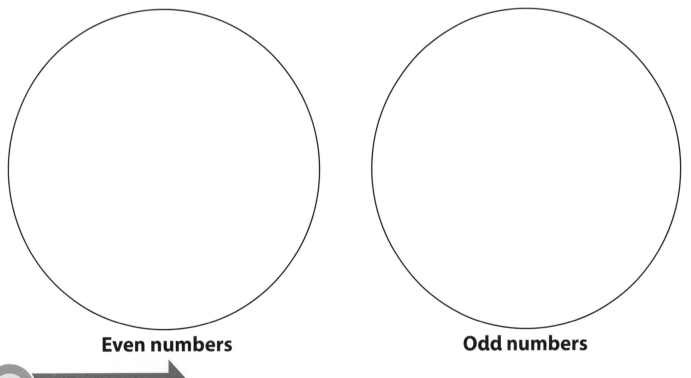

Even numbers   **Odd numbers**

Stretch zone

Write the biggest even number you know. _____

Write the smallest odd number you know. _____

3F Even and odd

Explore 1   Student Book 1, page 51

Follow these steps six times.

- Shuffle the digit cards.
- Turn over two cards.
- Use the digits to complete the number sentence.
- Is the answer even or odd?

digit cards 1–9

I turned over 5 and 4.

| 5 | 4 |

An example is shown below.

	5	+	4	=	9	

9 is an _____odd_____ number.

1   ☐ + ☐ = ☐   ☐ is an _____ number.

2   ☐ + ☐ = ☐   ☐ is an _____ number.

3   ☐ + ☐ = ☐   ☐ is an _____ number.

4   ☐ + ☐ = ☐   ☐ is an _____ number.

5   ☐ + ☐ = ☐   ☐ is an _____ number.

6   ☐ + ☐ = ☐   ☐ is an _____ number.

Stretch zone

True or false? If I add two odd numbers together, the total is always even. Explain your answer.

_____

_____

3F Even and odd

Explore 2 Student Book 1, page 52

Follow these steps two times.

- Is the start number odd or even?
- Count on from the start number in twos and then in fives. Write the sequences in the grids.

The first one is done for you.

Start number: 4   The start number is ____even____.

Count on in twos:

6	8	10	12	14	16	18	20

Count on in fives:

9	14	19	24	29	34	39	44

Start number: 7   The start number is _____.

Count on in twos:

Count on in fives:

Start number: 6   The start number is _____.

Count on in twos:

Count on in fives:

 Stretch zone

Talk to a partner or an adult about the patterns you see in the sequences.

3 Exploring numbers

Review

1 Draw a face next to each bubble to show how you feel about your learning.

comparing and ordering numbers

saying the number 1 more or 1 less

partitioning into tens and ones

odd and even numbers

2 Tell a partner about one thing you did really well in this unit.

3 Draw or write about things you found easy, challenging or really hard.

What work did you feel confident doing?

What work was challenging?

Is there any work you might need some extra help with?

4A Combining sets

Discover 1   Student Book 1, page 56

Follow these steps eight times.

- Shuffle the digit cards.
- Turn over two cards.
- Write the digits in an addition sentence.
- Find the total.

- digit cards 1–9

You could use counters (or other small objects) to help you find the totals.

An example is shown below.

3 + 7 = 10

1 ☐ + ☐ = ☐   5 ☐ + ☐ = ☐

2 ☐ + ☐ = ☐   6 ☐ + ☐ = ☐

3 ☐ + ☐ = ☐   7 ☐ + ☐ = ☐

4 ☐ + ☐ = ☐   8 ☐ + ☐ = ☐

Stretch zone

How many different ways can you get a total of 7 ...

... by adding two digits, 1–9?

_____

... by adding three digits, 1–9?

_____

Discover 2   Student Book 1, page 57

Write the missing numbers.

You could use counters or the number line to help you.

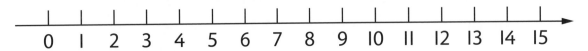

The first one is done for you.

$1 + 3 + \boxed{6} = 10$

1 $1 + \boxed{} + 3 = 10$

2 $\boxed{} + 3 + 1 = 10$

3 $\boxed{} + 4 + 1 = 10$

4 $5 + 1 + \boxed{} = 10$

5 $4 + \boxed{} + 2 = 10$

6 $\boxed{} + 4 + 4 = 10$

Find three different ways to make 15.

7 $\boxed{} + \boxed{} + \boxed{} = 15$

8 $\boxed{} + \boxed{} + \boxed{} = 15$

9 $\boxed{} + \boxed{} + \boxed{} = 15$

Stretch zone

How many different ways can you find to make 15 with three numbers? Write the different ways.

4A Combining sets

Explore 1 Student Book 1, page 58

Work out the missing numbers. Use the facts that you are given.

Use the number track to help you.

| 1 | 2 | 3 | 4 | 5 | 6 | 7 | 8 | 9 | 10 | 11 | 12 | 13 | 14 | 15 | 16 | 17 | 18 | 19 | 20 |

You can add numbers in any order. The answer is always the same.

The first one is done for you.

I know that $\boxed{7}$ + 4 = 11, so … $4 + \boxed{7} = 11$

1 I know that 5 + 7 = 12, so … $7 + \boxed{} = 12$

2 I know that $\boxed{}$ + 1 = 10, so … $1 + 9 = \boxed{}$

3 I know that 8 + 6 = 14, so … $6 + \boxed{} = 14$

4 I know that 3 + $\boxed{}$ = 11, so … $\boxed{} + 3 = 11$

5 I know that 5 + $\boxed{}$ = 14, so … $\boxed{} + 5 = 14$

6 I know that 1 + 11 = $\boxed{}$, so … $11 + \boxed{} = \boxed{}$

7 I know that 7 + 8 = $\boxed{}$, so … $8 + \boxed{} = \boxed{}$

8 I know that 7 + 9 = $\boxed{}$, so … $9 + \boxed{} = \boxed{}$

Stretch zone

Choose your own numbers to complete these number sentences.

$\boxed{} + 7 + \boxed{} = 20$ $\boxed{} + \boxed{} + 9 = 20$

$3 + \boxed{} + \boxed{} = 20$ $\boxed{} + \boxed{} + \boxed{} = 20$

4 Addition

4A Combining sets

Explore 2 Student Book 1, page 59

How many buttons in total? Complete the addition sentences.

You could use counters (or other small objects) to help you.

The first one is done for you.

3	+ 5	= 8

4

	+	=

1

	+	=

5

	+	=

2

	+	=

6

	+	=

3

	+	=

7

	+	=

Stretch zone

Write three different ways to make a total of 8.

_____ _____ _____

62

Explore 3   Student Book 1, page 60

Draw cubes next to each addition sentence to help you find the total.

You could use real cubes (or other small objects) in two different colours to help you.

The first one is done for you.

7 + 3 = | 10 |

1 7 + 4 =

2 8 + 4 =

3 4 + 8 =

4 9 + 6 =

5 9 + 7 =

Stretch zone

Write three different addition sentences that use three numbers.

☐ + ☐ + ☐ = ☐

☐ + ☐ + ☐ = ☐

☐ + ☐ + ☐ = ☐

Discover   Student Book 1, page 61

> • colouring pencils: two different colours

Follow these steps five times.

• Look at the first number in the addition. Colour that number of squares.

• Look at the second number in the addition. Colour that number of squares, in a different colour.

• Write the total.

An example is shown below.

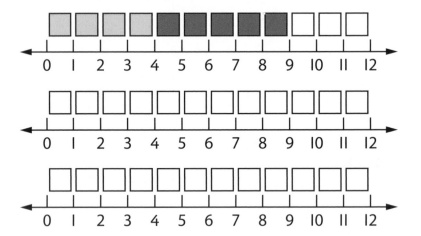

4 + 5 = $\boxed{9}$

1 1 + 8 = $\boxed{}$

2 8 + 2 = $\boxed{}$

3 3 + 5 = $\boxed{}$

4 6 + 6 = $\boxed{}$

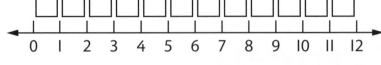

5 7 + 4 = $\boxed{}$

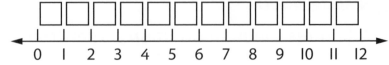

Stretch zone

What is the largest total you can make by adding two digits? $\boxed{}$

What is the smallest total? $\boxed{}$

Draw the jumps on the number line. Write the answer to each addition.
The first one is done for you.

$3 + 8 + 4 =$ 15

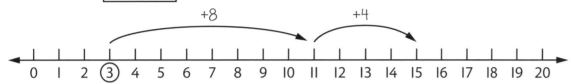

1 $8 + 3 + 4 =$ [ ]

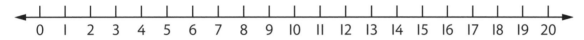

2 $4 + 3 + 8 =$ [ ]

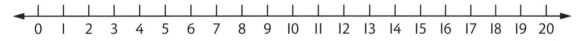

3 $7 + 7 + 6 =$ [ ]

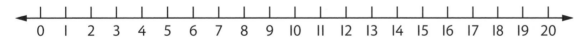

4 $7 + 7 + 5 =$ [ ]

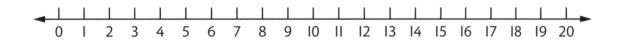

Stretch zone

Write three addition sentences that total 20. Each one should involve adding three numbers.

4 Addition

Discover Student Book 1, page 63

Use the number track to write the missing numbers.

| 1 | 2 | 3 | 4 | 5 | 6 | 7 | 8 | 9 | 10 | 11 | 12 | 13 | 14 | 15 | 16 | 17 | 18 | 19 | 20 |

The first one is done for you.

I start on 2. I count on 5. I land on [7].

1 I start on 4. I count on 5. I land on [ ].

2 I start on 6. I count on 5. I land on [ ].

3 I start on 10. I count on 3. I land on [ ].

4 I start on 10. I count on 1. I land on [ ].

5 I start on 10. I count on 6. I land on [ ].

6 I start on 15. I count on 4. I land on [ ].

7 I start on 16. I count on 4. I land on [ ].

8 I start on 18. I count on [ ]. I land on 20.

9 I start on 12. I count on [ ]. I land on 20.

10 I start on 14. I count on [ ]. I land on 20.

Stretch zone

I know that 16 + 4 = 20. This means that 20 − 4 = 16.

Write three more statements like this.

Explore 1   Student Book 1, page 64

Find the number bonds to 10 to help you complete each addition.

The first one is done for you.

$\boxed{2} + \boxed{8} + \boxed{3} = \boxed{13}$

3

$\boxed{} + \boxed{} + \boxed{} = \boxed{}$

1

$\boxed{} + \boxed{} + \boxed{} = \boxed{}$

4

$\boxed{} + \boxed{} + \boxed{} = \boxed{}$

2

$\boxed{} + \boxed{} + \boxed{} = \boxed{}$

5

$\boxed{} + \boxed{} + \boxed{} = \boxed{}$

Stretch zone

Make up your own additions. Write three numbers on the balloons. Write the addition sentences.

_____   _____

Explore 2   Student Book 1, page 65

Draw the jumps on the number line. Write the answer to each addition.

1 9 + 7 = ☐

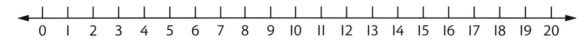

2 9 + 5 = ☐

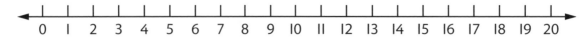

3 9 + 3 = ☐

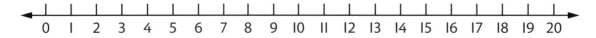

4 17 + 2 = ☐

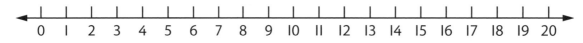

5 2 + 17 = ☐

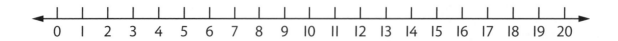

Stretch zone

Complete this addition sentence. Look at the number lines to help you.

☐ + ☐ + ☐ = 16

4D Bridging 10

Discover
Student Book 1, page 66

- colouring pencils: two different colours

Draw counters in the ten-frames to match the numbers.

Write the complete addition sentence.

The first one is done for you.

	6 + 7		6 + 7 = 13
1	8 + 4		
2	9 + 7		
3	9 + 8		
4	9 + 9		
5	8 + 6		
6	6 + 6		

Stretch zone

 How many number bonds to 16 can you find?

Draw your answers using ten-frames.

Explore   Student Book 1, page 67

Write the answers to these additions.

1   $9 + 6 =$ ☐

2   $9 + 7 =$ ☐

3   $9 + 3 =$ ☐

4   $9 + 9 =$ ☐

5   $8 + 6 =$ ☐

6   $8 + 7 =$ ☐

7   $8 + 3 =$ ☐

8   $8 + 9 =$ ☐

9   $7 + 6 =$ ☐

10   $7 + 7 =$ ☐

11   $7 + 3 =$ ☐

12   $7 + 9 =$ ☐

13   $6 + 5 =$ ☐

14   $6 + 6 =$ ☐

15   $5 + 7 =$ ☐

16   $5 + 8 =$ ☐

Stretch zone

What is a quick way to find the answer if the starting number is 9?

(For example: $9 + 5$.)

_____

_____

4E Addition word problems

Discover   Student Book 1, page 69

- colouring pencils: two different colours

Solve these addition stories. Draw counters in the ten-frames to help you.

1 Tanvi has 8 reading books. Her friend has 5 reading books. How many do they have altogether?

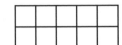   ☐ + ☐ = ☐

2 I have 6 blue building blocks and 6 red building blocks. How many do I have altogether?

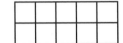   

3 I have 8 party hats. My mum buys 6 more. How many do I have now?

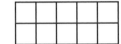   

4 There are 5 orange fish and 7 yellow fish in the fish tank. How many fish altogether?

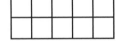   ☐ + ☐ = ☐

5 I have 6 grapes and 7 cherries. How many pieces of fruit do I have?

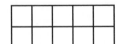   

Stretch zone

Write an addition story to match these ten-frames.

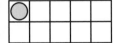

_____

_____

Explore   Student Book 1, page 70

Complete the addition stories to match these pictures. Write the addition sentences.

1

There are ☐ helicopters. There are ☐ airplanes.

There are ☐ aircraft altogether.   ☐ + ☐ = ☐

2

There are ☐ ducks in one pond. There are ☐ ducks in the other pond.

There are ☐ ducks altogether.   ☐ + ☐ = ☐

3

There are ☐ socks with stripes. There are ☐ socks with spots.

There are ☐ socks altogether.   ☐ + ☐ = ☐

4

There are ☐ ladybirds. There are ☐ butterflies.

There are ☐ insects altogether.   ☐ + ☐ = ☐

Stretch zone

 Draw a picture to match this addition sentence. Write the story.

5 + 7 = 12

Review

1 Draw a face next to each bubble to show how you feel about your learning.

adding small numbers

using number lines to add

using number bonds to solve additions

solving word problems

2 Tell a partner about one thing you did really well in this unit.

3 Draw or write about things you found easy, challenging or really hard.

What work did you feel confident doing?

What work was challenging?

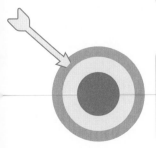

Is there any work you might need some extra help with?

5A Counting back

Discover   Student Book 1, page 74

Choose any numbers to count back. Complete the number sentences.

Use the number track to help you.

| 1 | 2 | 3 | 4 | 5 | 6 | 7 | 8 | 9 | 10 | 11 | 12 | 13 | 14 | 15 | 16 | 17 | 18 | 19 | 20 |

The first one is done for you.

$$10 - \boxed{0} = \boxed{10}$$

1   $10 - \boxed{} = \boxed{}$

2   $10 - \boxed{} = \boxed{}$

3   $10 - \boxed{} = \boxed{}$

4   $10 - \boxed{} = \boxed{}$

5   $15 - \boxed{} = \boxed{}$

6   $15 - \boxed{} = \boxed{}$

7   $15 - \boxed{} = \boxed{}$

8   $15 - \boxed{} = \boxed{}$

9   $20 - \boxed{} = \boxed{}$

10 $20 - \boxed{} = \boxed{}$

11 $20 - \boxed{} = \boxed{}$

12 $20 - \boxed{} = \boxed{}$

Stretch zone

Check your answers to questions 11 and 12 using addition.

For example: $20 - 12 = 8$. Check by calculating $12 + 8$. Is the answer 20?

_____   _____

5A Counting back

Explore Student Book 1, page 75

Choose any numbers to count back. Complete the number sentences.

Use the number track to help you.

| 1 | 2 | 3 | 4 | 5 | 6 | 7 | 8 | 9 | 10 | 11 | 12 | 13 | 14 | 15 | 16 | 17 | 18 | 19 | 20 |

The first one is done for you.

17 − [3] = [14]

1 20 − [　] = [　]

2 16 − [　] = [　]

3 7 − [　] = [　]

4 18 − [　] = [　]

5 9 − [　] = [　]

6 4 − [　] = [　]

7 17 − [　] = [　]

8 13 − [　] = [　]

9 10 − [　] = [　]

10 15 − [　] = [　]

11 19 − [　] = [　]

12 8 − [　] = [　]

Stretch zone

Check your answers to questions 11 and 12 using addition.

For example: 19 − 8 = 11. Check by calculating 11 + 8. Is the answer 19?

_____   _____

5 Subtraction and difference

75

5B Taking away

Discover   Student Book 1, page 76

Draw the jump on each number line to show the subtraction.

Complete the number sentences.

You could use buttons, counters or coins to help you.

The first one is done for you.

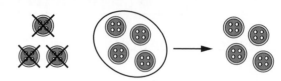

3 less than 7 is [4].

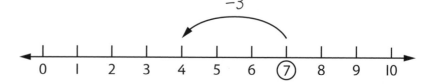

1 4 less than 9 is [ ].

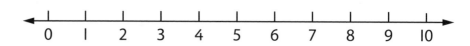

2 2 less than 7 is [ ].

3 5 less than 9 is [ ].

4 1 less than 8 is [ ].

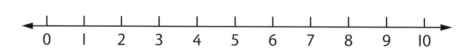

5 4 less than 6 is [ ].

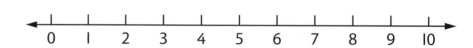

Stretch zone

Draw a number line to show the subtraction 13 – 5.

5B Taking away

Explore I   Student Book 1, page 77

Complete these subtraction calculations.

Use the number track to help you.

I	2	3	4	5	6	7	8	9	10

The first one is done for you.

I less than 10 is 9 .

I 2 less than 10 is [ ] .

2 3 less than 10 is [ ] .

3 2 less than 10 is [ ] .

4 2 less than 5 is [ ] .

5 3 less than 5 is [ ] .

6 4 less than 5 is [ ] .

7 2 less than 7 is [ ] .

8 3 less than 7 is [ ] .

9 4 less than 7 is [ ] .

10 [ ] less than [ ] = [ ]

11 [ ] less than [ ] = [ ]

12 [ ] less than [ ] = [ ]

Stretch zone

Write a sentence about the patterns you see in your answers.

You could ask an adult to help with the writing.

_____

_____

Explore 2   Student Book 1, page 78

Follow these steps four times.

- Shuffle the digit cards.

• digit cards 1–9

- Turn over a card. Write the number in the subtraction sentence.

- Draw the jump on the number line. Complete the subtraction sentence.

An example is shown below.

13 − $\boxed{3}$ = $\boxed{10}$
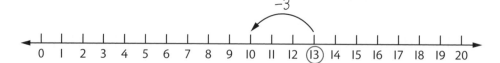

1   18 − $\boxed{}$ = $\boxed{}$

2   12 − $\boxed{}$ = $\boxed{}$
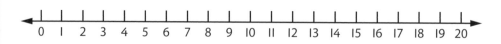

3   11 − $\boxed{}$ = $\boxed{}$
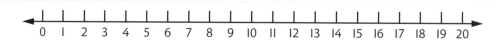

4   20 − $\boxed{}$ = $\boxed{}$

Stretch zone

Turn over three digit cards. Use the digits to write three subtraction questions. Each question must have a 2-digit number and a 1-digit number, for example, 42 − 5. Work out the answers.

5B Taking away

- 18 cubes (or other small objects)
- 0–9 digit cards (you could make your own)

- Place the cubes on a table.
- Pick a card.
- Take away the number of cubes on the card.
- Write the subtraction.

Repeat five more times.

An example is shown below.

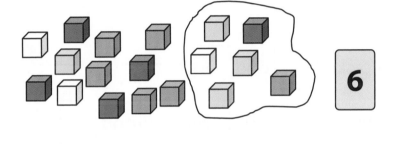

$18 - \boxed{6} = \boxed{12}$

1   $18 - \boxed{} = \boxed{}$

2   $18 - \boxed{} = \boxed{}$

3   $18 - \boxed{} = \boxed{}$

4   $18 - \boxed{} = \boxed{}$

5   $18 - \boxed{} = \boxed{}$

6   $18 - \boxed{} = \boxed{}$

Stretch zone

Write an easy subtraction with the answer 12. $\boxed{} - \boxed{} = 12$

Write a hard subtraction with the answer 12. $\boxed{} - \boxed{} = 12$

5 Subtraction and difference

5C Finding the difference

Discover   Student Book 1, page 80

- Pick two cards.
- Draw a jump on the number line to find the difference.

Repeat three more times.

An example is shown below.

I picked [5] and [8].

The difference between [5] and [8] is [3].

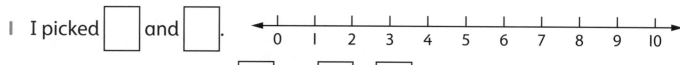

1 I picked [ ] and [ ].

The difference between [ ] and [ ] is [ ].

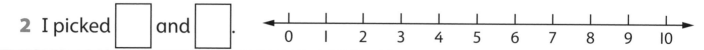

2 I picked [ ] and [ ].

The difference between [ ] and [ ] is [ ].

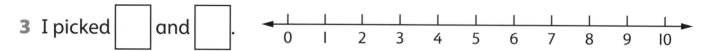

3 I picked [ ] and [ ].

The difference between [ ] and [ ] is [ ].

Stretch zone

Write three pairs of numbers that each have a difference of 3.

_____   _____   _____

5C Finding the difference

Complete the number puzzles.

The number on top is the difference between the two numbers below.

Use the number track to help you.

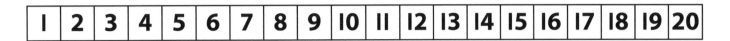

The first one is done for you.

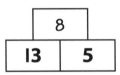

3
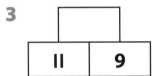

6
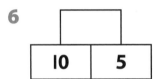

1
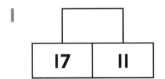

4
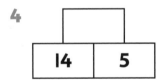

7
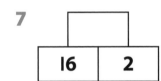

2
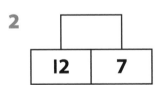

5
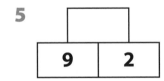

8
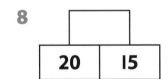

Stretch zone

Here is a bigger puzzle. Can you complete it in the same way?

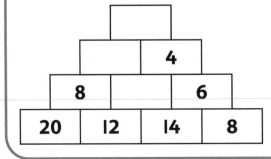

5 Subtraction and difference

81

5D Subtraction word problems

Complete the subtraction stories to match these pictures.

Write the subtraction sentences.

1 There are ⬜ eggs.

   ⬜ eggs are broken.

   ⬜ unbroken eggs are left.

   ⬜ – ⬜ = ⬜

2 Somchai has ⬜ bananas.

   He eats ⬜ bananas.

   He has ⬜ bananas left.

   ⬜ – ⬜ = ⬜

3 We have ⬜ apples.

   We eat ⬜ apples.

   We have ⬜ apples left.

   ⬜ – ⬜ = ⬜

4 ⬜ frogs are on a lily pad.

   ⬜ frogs jump off.

   ⬜ frogs are left on the lily pad.

   ⬜ – ⬜ = ⬜

Stretch zone

Draw a picture to match this subtraction sentence.

15 – 6 = 9

5D Subtraction word problems

Write the subtraction sentences to match these stories.

1   There are 15 birds in a tree. Then 4 birds fly away. How many are left?

☐ − ☐ = ☐

2   I have 9 cherries. Then I eat 4 cherries. How many are left?

☐ − ☐ = ☐

3   I have 14 pencils. Then my friend borrows 4 pencils. How many do I have left?

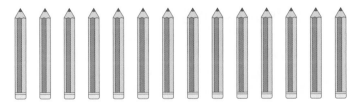

☐ − ☐ = ☐

4   I have 12 coins. Then I give my brother 2 coins. How many do I have left?

☐ − ☐ = ☐

5   8 babies are playing. 7 babies go to sleep. How many are still playing?

☐ − ☐ = ☐

Stretch zone

Write your own subtraction story for this picture.

Review

1 Draw a face next to each bubble to show how you feel about your learning.

using number lines to subtract

using number bonds to solve subtractions

finding differences

solving word problems

2 Tell a partner about one thing you did really well in this unit.

3 Draw or write about things you found easy, challenging or really hard.

What work did you feel confident doing?

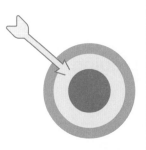

What work was challenging?

Is there any work you might need some extra help with?

6A Equal sharing

Discover
Student Book 1, page 87

- colouring pencils: blue and red

Which of these numbers can you share equally into two groups?

- If you **can** share the number equally, colour it blue.

- If you **cannot** share the number equally, colour it red.

You could use counters or cubes to help you.

21	22	23	24	25
26	27	28	29	30
31	32	33	34	35
36	37	38	39	40

- Describe to a friend or an adult the patterns in the table.

Stretch zone

What is the difference between odd and even numbers?

_____

_____

6A Equal sharing

Explore 1   Student Book 1, page 88

Can you share each set of objects equally between 2, 3 and 4 people?

> You could use buttons or coins (or other small objects) and try to share them into equal groups.

The first one is done for you.

	Objects		2 people	3 people	4 people
	8 toy cars		yes	no	yes
1	12 cakes				
2	9 balloons				
3	5 bananas				
4	10 eggs				

Stretch zone

Can you find a number of objects that you can share equally into groups of 2, 3 and 4 with none left over?

6A Equal sharing

Explore 2   Student Book 1, page 89

Draw pictures to solve these problems.   Keep your pictures very simple!

1 Share 12 cherries equally between 3 children.

2 Share 12 marbles equally between 4 children.

3 Share 16 sweets equally between 2 children.

4 Share 12 yo-yos equally between 6 children.

5 Share 10 apples equally between 5 children.

Stretch zone

How many different ways are there of sharing 24 objects equally?

6B Grouping

Discover   Student Book 1, page 90

The pieces of fruit need to go into bags.

Draw pictures to show how to fill the bags.

Write how many bags you can fill.

The first one is done for you.

Keep your pictures very simple!

Fruit	Number in each bag	Bag 1	Bag 2	Bag 3	Bag 4	
(apples)	7	(apples)	(apples)			I can fill 2 bags of apples.
(oranges)	4					I can fill bags of oranges.
(cherries)	6					I can fill bags of cherries.
(bananas)	3					I can fill bags of bananas.

Stretch zone

 Write and draw your own grouping problem.

6B Grouping

Write the missing numbers. Then complete the sentences.

The first one is done for you.

	If there are …	There must be …	'How many' sentence
I cat has 2 eyes.	8 eyes	4 cats	__4__ groups of __2__ equals __8__
1 — I boy has 2 legs.	10 legs	☐ boys	____ groups of ____ equals ____
2 — I ostrich has 2 legs.	12 legs	☐ ostriches	____ groups of ____ equals ____
3 — I bird has 2 wings.	6 wings	☐ birds	____ groups of ____ equals ____
4 — I pair has 2 socks.	14 socks	☐ pairs of socks	____ groups of ____ equals ____

Stretch zone

Draw a picture and write your own 'How many' question.

_____

_____

Write a 'how many' sentence to answer your question. ____ groups of ____ equals ____

Discover   Student Book 1, page 92

Write four different number sentences about this picture.

Some should involve addition. Some should involve 'groups of'.

1 _____

2 _____

3 _____

4 _____

Stretch zone

Use the picture to write a word problem for a friend to solve.

You must be able to solve the problem yourself.

_____

_____

_____

_____

6C Repeated addition

Explore   Student Book 1, page 93

Draw counters to show the following calculations.

The first one is done for you.

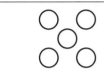

$5 + 5 =$ [10]

$2 \times 5 =$ [10]

1

$6 + 6 =$ [ ]

$2 \times 6 =$ [ ]

3

$8 + 8 =$ [ ]

$2 \times 8 =$ [ ]

2

$7 + 7 =$ [ ]

$2 \times 7 =$ [ ]

4

$9 + 9 =$ [ ]

$2 \times 9 =$ [ ]

Stretch zone

Draw pictures to show different ways of making 16 using equal groups.

How do you know you have all the different ways? Discuss with a partner or an adult.

Discover　　　Student Book 1, page 94

For each picture, write the missing numbers.

Write a repeated addition sentence. Then write a multiplication sentence.

The first one is done for you.

| 3 | bunches of cherries. | 4 | in each bunch.

$\underline{\quad 4 + 4 + 4 = 12 \quad}$　　　$\underline{\quad 3 \times 4 = 12 \quad}$

1　　☐ bunches of bananas. ☐ in each bunch.

_____　　_____

2　☐ pots of pencils. ☐ in each pot.

_____　　_____

3　　☐ piles of books. ☐ in each pile.

_____　　_____

4　☐ bunches of balloons. ☐ in each bunch.

_____　　_____

Stretch zone

Can you think of two different multiplication problems with the answer 12? Draw and write them.

6D Multiplication word problems

Explore   Student Book 1, page 95

Solve these word problems. Draw pictures if you like.

You could use counters (or other small objects) to help you.

1 My little brother has 2 tricycles. Each has 3 wheels. How many wheels altogether?

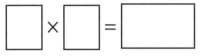

2 There are 5 birds in a tree. Each has 2 legs. How many legs altogether?

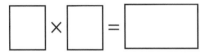

3 There are 3 spiders on a wall. Each has 8 legs. How many legs altogether?

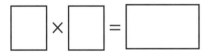

4 In a classroom there are 5 tables. Each has 4 legs. How many legs altogether?

$\boxed{} \times \boxed{} = \boxed{}$

Stretch zone

On a farm there are children and goats. I can see 24 legs.

How many children and goats could there be?

Can you find more than one solution? Write your answers.

93

Review

1 Draw a face next to each bubble to show how you feel about your learning.

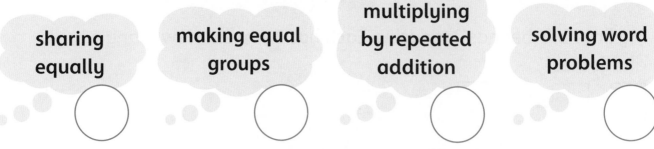

sharing equally

making equal groups

multiplying by repeated addition

solving word problems

2 Tell a partner about one thing you did really well in this unit.

3 Draw or write about things you found easy, challenging or really hard.

What work did you feel confident doing?

What work was challenging?

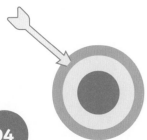

Is there any work you might need some extra help with?

7A Doubles and halves

Discover Student Book 1, page 99

Colour squares in each tower to match the shaded squares.

Count all the shaded squares to find the doubles.

The first one is done for you.

 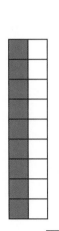 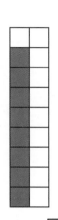

Double 6 is ⎢ 12 ⎥.   **1** Double 9 is ⎢   ⎥.   **2** Double 8 is ⎢   ⎥.

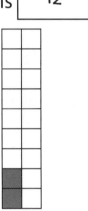 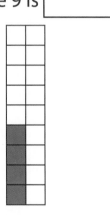 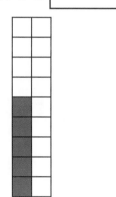

3 Double 2 is ⎢   ⎥.   **4** Double 4 is ⎢   ⎥.   **5** Double 5 is ⎢   ⎥.

 Stretch zone

Make up two doubles of your own.

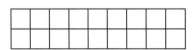   

Double ⎢ ⎥ is ⎢   ⎥.   Double ⎢ ⎥ is ⎢   ⎥.

Explore   Student Book 1, page 100

Draw an oval around half the objects in each group.
Complete the number sentences.

The first one is done for you.

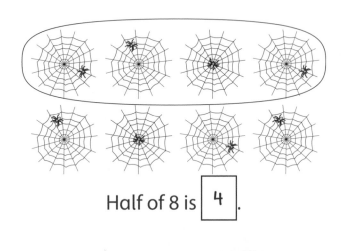

Half of 8 is 4.

3

Half of 4 is ☐.

1

Half of 6 is ☐.

4

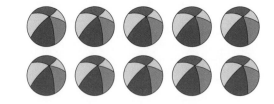

Half of 10 is ☐.

2

Half of 2 is ☐.

5

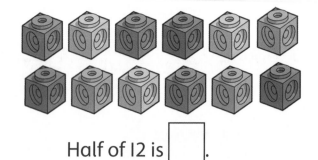

Half of 12 is ☐.

Stretch zone

Can you find numbers to complete this sentence?

Half of ☐ is the same as double ☐.

7B Halves

Discover   Student Book 1, page 101

- colouring pencils
- a ruler

Find two different ways to divide each shape in half.

Colour one half of the shape.

The first one is done for you.

1

2

Find half of each of these numbers. Write the number.

Draw a picture to show the two halves.

The first one is done for you.

You could use counters (or other small objects) to help you.

	Number	Half ($\frac{1}{2}$)	Picture
	14	7	
3	8		
4	12		
5	16		

What is the largest number that you can find half of? Write the number and the answer.

Half of ⬚ is ⬚ .

7B Halves

Explore   Student Book 1, page 102

• colouring pencils

I Draw three different squares or rectangles on this grid. Colour half of each shape.

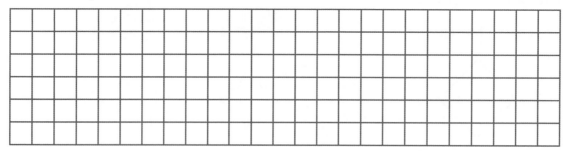

Find half of each number shown below . Draw circles to show the two halves.

The first one is done for you.

$\frac{1}{2}$ of 8 = [4].

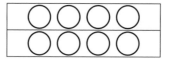

2 $\frac{1}{2}$ of 10 = [ ]

3 $\frac{1}{2}$ of 12 = [ ]

4 $\frac{1}{2}$ of 14 = [ ]

5 $\frac{1}{2}$ of 16 = [ ]

Stretch zone

What do you notice about the answers?

_____

_____

Discover   Student Book 1, page 103

- colouring pencils   • a ruler

There are different ways to colour a quarter ($\frac{1}{4}$) of a shape.

In each of these pictures $\frac{1}{4}$ of the shape is coloured.

Find six different ways to colour $\frac{1}{4}$ of a square.

1    **2**    **3**

4    **5**    **6**

Stretch zone

Find an easy way to colour $\frac{1}{4}$ of this rectangle.

Find a harder way to colour $\frac{1}{4}$ of this rectangle.

7 Fractions

Explore   Student Book 1, page 104

Find one quarter of each quantity of counters.

Draw counters inside the quarters of each circle to help you.

The first one is done for you.

12 counters

$\frac{1}{4}$ of 12 = $\boxed{3}$

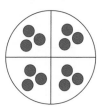

1 8 counters

$\frac{1}{4}$ of 8 = $\boxed{}$

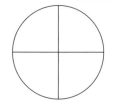

3 24 counters

$\frac{1}{4}$ of 24 = $\boxed{}$

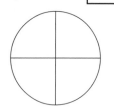

5 16 counters

$\frac{1}{4}$ of 16 = $\boxed{}$

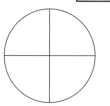

2 4 counters

$\frac{1}{4}$ of 4 = $\boxed{}$

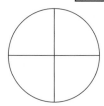

4 20 counters

$\frac{1}{4}$ of 20 = $\boxed{}$

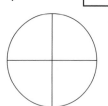

6 28 counters

$\frac{1}{4}$ of 28 = $\boxed{}$

Stretch zone

Choose your own number: $\boxed{}$. Draw a picture to show $\frac{1}{4}$.

$\frac{1}{4}$ of $\boxed{}$ = $\boxed{}$

7 Fractions

Review

1 Draw a face next to each bubble to show how you feel about your learning.

> doubling and halving numbers

> finding half of a number or shape

> finding quarter of a number or shape

2 Tell a partner about one thing you did really well in this unit.

3 Draw or write about things you found easy, challenging or really hard.

What work did you feel confident doing?

What work was challenging?

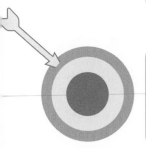

Is there any work you might need some extra help with?

8A Length

Discover   Student Book 1, page 108

Find six objects in your home or classroom.

Draw them in order, from shortest to longest.

When you compare lengths, make sure the objects are lined up correctly. This makes it easy to see which object is longer.

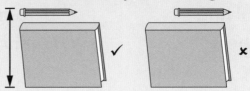

Shortest

1	2	3
4	5	6

Longest

Stretch zone

Use finger widths to measure the shortest object you found.

The _____ is about [   ]

finger widths long.

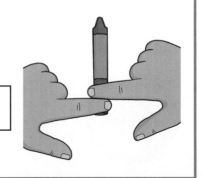

8A Length

Explore 1   Student Book 1, page 109

A hand span is the distance across your hand when you spread your fingers.

A pace is the length of one step forward.

Find objects in your home, classroom or outdoors to match the lengths in the table.

Length	Object
I hand span	
2 hand spans	
3 hand spans	
4 paces	
3 paces	
6 paces	

Stretch zone

Can you find two objects that are the same length?

Check by measuring them. What will you use to measure?

Complete this sentence.

_____ and _____ are both

_____ long.

8 Length, mass and capacity

Explore 2   Student Book 1, page 110

- lots of identical small objects, such as cubes, paper clips or dried pasta pieces

Choose four objects in your home or classroom. Measure their length using identical small objects. Complete the table.

An example is shown in the table.

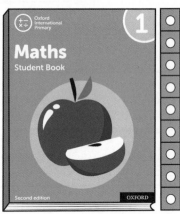

Object	Measurement
length of book	9 cubes

Stretch zone

Complete these sentences about the objects you found.

The longest object is _____. It measures _____.

The shortest object is _____. It measures _____.

Explore 3   Student Book 1, page 111

Use this ruler to measure the objects in the picture.

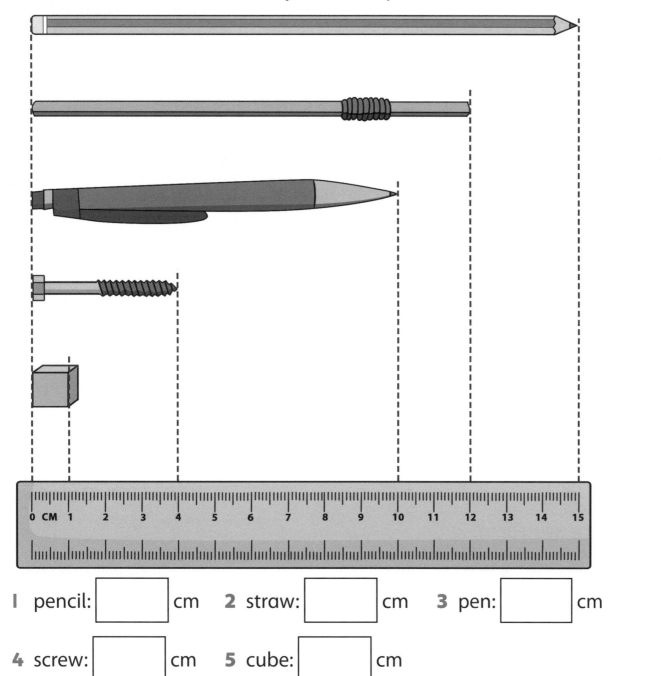

I pencil: ☐ cm   **2** straw: ☐ cm   **3** pen: ☐ cm

4 screw: ☐ cm   **5** cube: ☐ cm

Stretch zone ➤

Find an object in your home that is longer than the screw but shorter than the pen.

Write its name and length in cm. _____

Discover Student Book 1, page 112

Find six objects around your home or classroom.

Compare the masses of pairs of objects by holding one in each hand. Which one feels heavier?

Draw the objects in order, from lightest to heaviest.

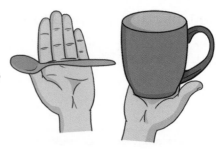

Lightest

1	2	3
4	5	6

Heaviest

Complete the sentences.

The ———————————————————————— feels the heaviest.

The ———————————————————————— feels the lightest.

  **Stretch zone**

 Write or draw:

- three things that are heavier than you
- three things that are lighter than you
- one thing with about the same mass as you.

8B Mass

- weighing scales
- things to weigh, for example different fruits

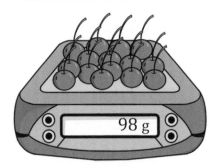

98 g

Use the scales to find things with a mass of about 100 grams.
Complete the list below.

The first one is done for you.

Things with a mass of about 100 grams:

- ___15 cherries_____

- _____

- _____

- _____

- _____

Stretch zone

Find objects to help you complete these sentences.

The heaviest object I can hold is _____.

The lightest object in my home is _____.

8B Mass

Explore 2   Student Book 1, page 114

- weighing scales
- things to weigh

Find six different objects. Use the scales to find the mass of each object. Complete the table.

Object	Mass in grams

Complete these sentences.

The heaviest object is _____ . Its mass is [ ] g.

The lightest object is _____ . Its mass is [ ] g.

Stretch zone

Find three objects with a mass of less than 50 g. Write them here.

Object: _____   Mass: [ ] g

Object: _____   Mass: [ ] g

Object: _____   Mass: [ ] g

8C Estimating capacity

Discover   Student Book 1, page 115

- some large dried chickpeas (or dried beans)
- four different small containers (jars or tubs)
- a small cup

Start with the smallest container.

- Write the type of container in the table.
- Estimate how many small cups of chickpeas will fill it. Write your estimate.
- Fill the container with chickpeas. Count the number of cups and write the actual number.
- Use what you learned from this estimate to help with your next estimate. Repeat for the other three containers.

Container	Estimate	Actual number

Stretch zone

Find a container that you think has a smaller capacity than the other four.

_____

Check by filling it with cups of chickpeas. Were you right? _____

8C Estimating capacity

Explore 1   Student Book 1, page 116

Find six different types of object that you can hold lots of in your hands.

You could use buttons, cherries,
dried beans, marbles, …

- Write the name of the object.
- Estimate how many you can fit in one hand, and how many in two hands.
- Try it out and write the actual numbers.

Repeat for the other five types of object.

Object	1 hand		2 hands	
	Estimate	Actual	Estimate	Actual

Stretch zone

Which objects were easiest to estimate? Why?

_____

_____

_____

8C Estimating capacity

Explore 2   Student Book 1, page 117

- a measuring jug
- water
- five small containers

Use the measuring jug to measure the capacity of each container.

Complete the table.

Container	Capacity in millilitres (ml)

Complete these sentences.

The _____ holds the most. It holds [   ] ml.

The _____ holds the least. It holds [   ] ml.

Stretch zone

Find a container with a capacity of about 500 ml. _____

Check the capacity using the measuring jug. Were you right? _____

Review

I Draw a face next to each bubble to show how you feel about your learning.

estimating, comparing and measuring lengths ◯

estimating, comparing and measuring masses ◯

estimating, comparing and measuring capacities ◯

2 Tell a partner about one thing you did really well in this unit.

3 Draw or write about things you found easy, challenging or really hard.

What work did you feel confident doing?

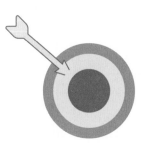

What work was challenging?

Is there any work you might need some extra help with?

9A Money amounts

Discover
Student Book 1, page 121

Think of four small items you can buy in a shop. Write them in the table.

Ask an adult how much each item costs. Write the costs in the table.

Write the coins you could use to buy each item.

You could use real coins to help you.

An example is shown in the table.

The example uses cents, but use coins from your country.

Item	Cost	Coins
mango	24¢	10¢ + 10¢ + 1¢ + 1¢ + 1¢ + 1¢

Stretch zone

Choose an item from the table. _____

Use the fewest coins possible to pay for the item. Write the coins here.

_____

Use the most coins possible to pay for the item. Write the coins here.

_____

9A Money amounts

Explore 1 — Student Book 1, page 122

Which currency are you using?

I am using _____.

Find three different ways of making each amount.
Write the coins.

You could use real coins to help you.

Currency means the money that people use in your country.

An example is shown in the table.

Amount	First way	Second way	Third way
20	20¢	10¢ + 5¢ + 5¢	5¢ + 5¢ + 5¢ + 5¢
22			
18			
10			
15			

Stretch zone

Can you make 25 using exactly three coins? Circle your answer.

Yes.   Write the coins that you used. _____

No.   Why not? _____

9A Money amounts

Explore 2   Student Book 1, page 123

Use the coins 1¢, 5¢, 50¢ and $1.

Make five amounts less than 50¢.
Write or draw the amounts in each box.

Remember, there are 100¢ in $1.

1

2

3

4

5

Make five amounts greater than 50¢. Write or draw the amounts in each box.

6

7

8

9

10

Stretch zone

 Make $1.25 in three different ways.

Write or draw the coins.

Remember: you can only use 1¢, 5¢, 50¢ and $1 coins.

Discover Student Book 1, page 124

Think of six toys. Write them in the table.

Ask an adult how much each toy costs. Write the costs in the table.

How can you use notes and coins to pay for each toy? Draw or write your answer in the table.

Toy	Cost	Notes and coins

Stretch zone

Show three different ways of paying for the most expensive toy in the table.

Explore   Student Book 1, page 125

My cousin in the UK uses pounds and pence.

Show eight different ways to make 20p.

You can use these coins: 1p, 2p, 5p, 10p and 20p.

An example is shown in the table.

5p + 5p + 10p

Stretch zone

You cannot use more than five coins. You can only use 1p, 2p, 5p and 10p coins.

How many different ways can you make 20p? Write the different ways.

Review

I Draw a face next to each bubble to show how you feel about your learning.

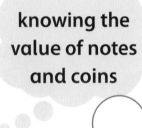

recognising notes and coins

knowing the value of notes and coins

making amounts using coins

2 Tell a partner about one thing you did really well in this unit.

3 Draw or write about things you found easy, challenging or really hard.

What work did you feel confident doing?

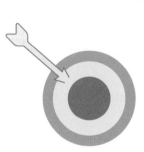

What work was challenging?

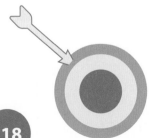

Is there any work you might need some extra help with?

Discover   Student Book 1, page 129

Write or draw what you do at each time of day.

 in the morning	
 at midday	
 in the afternoon	
 in the evening	
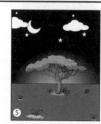 at night	

Stretch zone

Explain how you know when it is morning.

_____

Explain how you know when it is night.

_____

Explore   Student Book 1, page 130

- Write the correct word under each picture.
- Put the events in the correct order by writing 1, 2, 3 or 4 in each box.

morning   afternoon   evening   night

| | I eat dinner. | | I am asleep. |

_____   _____

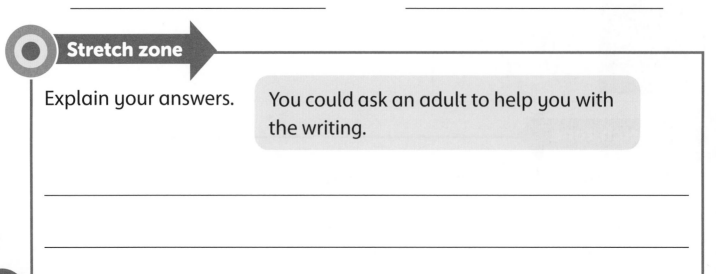

| | I wake up. | | I eat lunch. |

_____   _____

Stretch zone

Explain your answers.   You could ask an adult to help you with the writing.

_____

_____

_____

10B Days of the week

Monday	Tuesday	Wednesday	Thursday	Friday	Saturday	Sunday
		1	2	3	4	5
6	7	8	9	10	11	12
13	14	15	16	17	18	19
20	21	22	23	24	25	26
27	28	29	30	31		

Use this calendar to complete the tables.

	Day	Number of days in this month
1	Monday	
2	Tuesday	
3	Wednesday	
4	Thursday	

	Day	Number of days in this month
5	Friday	
6	Saturday	
7	Sunday	

Stretch zone

If 1 January is a Saturday, 8 January must be a _____.

If 6 January is a Thursday, 13 January must be a _____.

If 1 February is a Tuesday, 11 February must be a _____.

Explore   Student Book 1, page 132

Do you know the days of the week?

Draw something you do on each of these days.

WEEK	
Monday	
Tuesday	
Wednesday	
Thursday	
Friday	
Saturday	
Sunday	

Tuesday

Friday

Sunday

 Stretch zone

True or false?

There are always four Mondays in a month.   _____

Some months only have three Fridays.   _____

10C Telling the time

Discover   Student Book 1, page 133

Draw a picture to show what you do at these times of the day. Draw the hands on the clocks to show each time.

1	9 o'clock in the morning		
2	12 o'clock noon		
3	3 o'clock in the afternoon		
4	6 o'clock in the evening		

Stretch zone

How do you know what time this is?

_____

_____

10 Time

123

10C Telling the time

Explore 1   Student Book 1, page 134

Write the time shown on each clock. The first one is done for you.

half past 2

4
_____

1
_____

5
_____

2
_____

6
_____

3
_____

7
_____

Stretch zone

Draw hands on the clock to show your favourite time of day. Explain why it is your favourite time.

_____

_____

Explore 2   Student Book 1, page 135

Draw hands on each clock to show the times you do these things.

Then write each time.

wake up	(clock face)	
arrive at school	(clock face)	
lunch time	(clock face)	
end of school	(clock face)	

Stretch zone

Draw an easy time to say. Write the time.

_____

Draw a difficult time to say. Write the time.

_____

Discover   Student Book 1, page 136

- a stopwatch or timer

Do each activity in the table. Ask a partner or an adult to time you.

Before you look at the stopwatch, estimate how long it took. Write your estimate.

Look at the stopwatch. Write the actual time in minutes and seconds.

Activity	Estimated time	Actual time
write your full name 5 times		
spell out your full name		
touch your head and then your knee 10 times		
say the alphabet		
count to 100		

Stretch zone

How can you estimate one minute?

10D Measuring time

Explore   Student Book 1, page 137

Think of activities that take the amounts of time shown.

Write at least one activity for each amount of time.

Time	Activity
about I second	
about I0 seconds	
about 30 seconds	
about I minute	
about I0 minutes	
about 30 minutes	

Stretch zone

What is the longest time you have ever spent on one activity?

What was the activity? _____

How long did you spend on it? _____

Review

I Draw a face next to each bubble to show how you feel about your learning.

ordering
daily events

days of the
week

telling the time
to the hour and
half past

estimating and
measuring in
seconds and
minutes

 2 Tell a partner about one thing you did really well in this unit.

3 Draw or write about things you found easy, challenging or really hard.

What work did you feel confident doing?

What work was challenging?

Is there any work you might need some extra help with?

11A 2D shapes

Discover   Student Book 1, page 141

Draw a shape on each geoboard. Try to make all your shapes different.

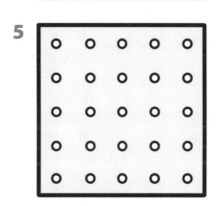

Stretch zone

Choose one of your shapes. Write the number of the shape here. _____

Write a sentence to describe the shape.

_____

_____

Explore 1   Student Book 1, page 142

Follow the instructions to draw shapes.
There are lots of possible shapes you can draw.

1 Draw a three-sided shape with two
   equal sides.

2 Draw a four-sided shape with one
   curved side and three straight sides.

3 Draw a shape with five sides.

4 Draw a shape with four corners and
   two equal sides.

5 Draw a round shape.

Stretch zone

Do you know the names of any of the shapes you have drawn? Write the
names next to the shapes.

11A 2D shapes

Explore 2   Student Book 1, page 143

Look around your home or classroom. You will see lots of 2D shapes.

Draw each of these shapes. Write the name of an object that matches the shape.

Shape	Drawing	Object
a circle		
a square		
a rectangle		
a different rectangle		
a triangle		

Stretch zone

What is the same about the two rectangles?

_____

What is different?

_____

Discover   Student Book 1, page 144

- building blocks

If you do not have building blocks, you could use boxes or tins.

Follow the instructions to build two towers.

Draw a picture of each tower. Or take photographs and stick them on the page.

You could ask an adult to help.

Build a tower using cuboids and cubes.

Build a tower using cylinders and cuboids.

Stretch zone

Build the tallest tower you can. How many blocks did you use?

11B 3D shapes

Explore 1   Student Book 1, page 145

Find three different 3D shapes in your home or classroom.

Draw each shape in the table. Or take photographs and stick them on the page.

You could ask an adult to help.

Find one edge, one face and one vertex on each shape. Draw lines from the words to show where these are on your shapes.

My shapes	Properties
	edge
	face
	vertex
	edge
	face
	vertex
	edge
	face
	vertex

Stretch zone

Can you name any of your shapes?
Write the names next to your drawings.

Explore 2   Student Book 1, page 146

- building blocks

If you do not have building blocks, you could use boxes and tins.

Make a model using lots of different-shaped blocks.

Draw your model. Or take a photograph and stick it on the page.

You could ask an adult to help.

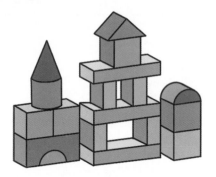

 Stretch zone

Write a sentence to describe your model.

Use some of these words:
edge, face, vertex, cube, cuboid, cone, cylinder, curved, straight, flat.

_____

_____

Explore 3   Student Book 1, page 147

Draw an arrow to match each shape to the correct name.

Some names will have more than one arrow.

1

                    cube

2

                    cuboid

3

                    sphere

4

                    square-based pyramid

5

                    triangular-based pyramid

6

Stretch zone

Compare the two pyramids. What is the same?

_____

What is different?

_____

Discover   Student Book 1, page 148

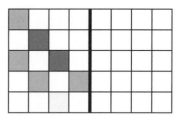

- colouring pencils

Play this game three times with a partner.

- Player I colours some squares on one side of the mirror line.

- Player 2 colours squares on the other side of the mirror line to make a symmetrical pattern.

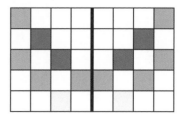

> Remember: a symmetrical pattern is exactly the same on each side of the mirror line.

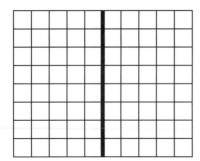   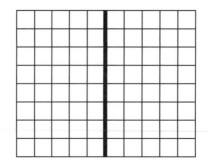   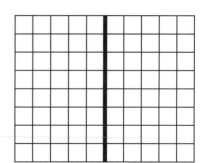

Stretch zone

Draw a simple symmetrical pattern.

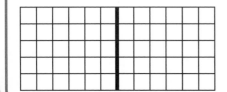

Draw a complicated symmetrical pattern.

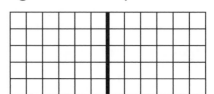

11C Symmetry

Explore 1   Student Book 1, page 149

This photograph shows an example of symmetry in nature.

Find four more pictures on the Internet or in magazines that show examples of symmetry.

Stick them in the boxes.

1

2

3

4

Stretch zone

Choose one of your pictures. Write the number here: _____

Explain how you know it is symmetrical.

_____

Explore 2   Student Book 1, page 150

• colouring pencils

Design your own symmetrical pattern.

Line of symmetry

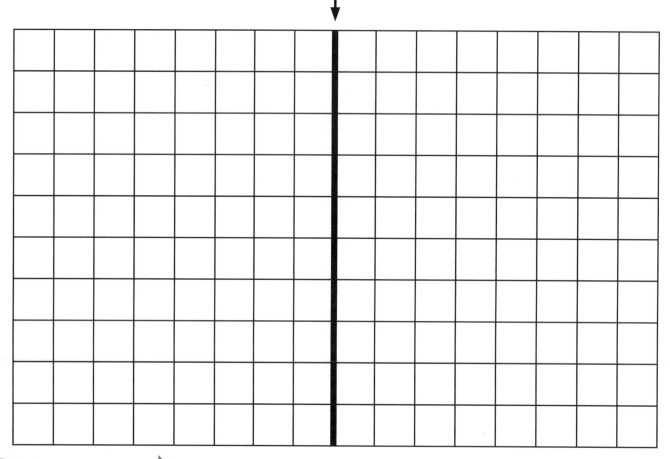

 Stretch zone

Design a pattern that is not symmetrical.

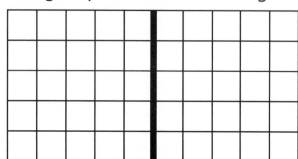

11D Position and movement

Talk to someone about what you see in this picture. Use some of these words:

next to   *in front of*   behind   **on top of**

above   under   below   between

Write two sentences about the picture.

You could ask an adult to help with the writing.

_____

_____

_____

_____

 **Stretch zone**

Draw a rectangle **next to** a triangle.	Draw a square **in front of** a circle.	Draw a triangle **on top of** a square.

11 Geometry

Explore   Student Book 1, page 152

Draw a picture next to each word or phrase.

Use the phrase to describe your drawing to a partner or an adult.

behind	
on top of	
underneath	
first	
last	

Stretch zone

 Draw a picture and write a description using three of the phrases.

11E Turns

Discover   Student Book 1, page 153

Cut out a simple shape from cardboard.

Draw your shape in the first row of the table.

Follow the instructions to turn the shape and draw how it looks each time.

Instruction	Example	My shape
start		
$\frac{1}{2}$ turn clockwise		
$\frac{1}{4}$ turn clockwise		
$\frac{3}{4}$ turn clockwise		

Stretch zone

If I turn my shape a $\frac{1}{4}$ turn anti-clockwise, it will look the same as if I turn it a [   ] turn clockwise.

Explore   Student Book 1, page 154

Look at the shape in each row.
Follow the instruction and draw the new shape.

Shape	Instruction	New shape
➡	quarter turn clockwise	
➡	quarter turn anti-clockwise	
➡	half turn	
⬌	half turn	
⬌	quarter turn anti-clockwise	
⬌	quarter turn clockwise	

Stretch zone

 Design a more complicated shape.
Draw some $\frac{1}{4}$, $\frac{1}{2}$ and $\frac{3}{4}$ turns.

11 Geometry

Review

1 Draw a face next to each bubble to show how you feel about your learning.

recognising and naming 2D and 3D shapes ◯

line symmetry ◯

describing position ◯

making whole, half, quarter and three-quarter turns ◯

2 Tell a partner about one thing you did really well in this unit.

3 Draw or write about things you found easy, challenging or really hard.

What work did you feel confident doing?

What work was challenging?

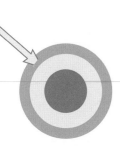
Is there any work you might need some extra help with?

Discover   Student Book 1, page 158

This pictogram tells us how many of each type of pizza a restaurant sold last Friday.

Pizzas sold on Friday

Type of pizza	Number sold					
margherita	🍕	🍕	🍕	🍕	🍕	🍕
mushroom	🍕	🍕	🍕	🍕		
four cheese	🍕	🍕				
chicken	🍕	🍕	🍕	🍕	🍕	

Key: 🍕 = 1 pizza

Use the pictogram to complete this table.

Pizzas sold on Friday

	Type of pizza	Number sold
1	margherita	
2	mushroom	
3	four cheese	
4	chicken	

Stretch zone

Write two things the pictogram tells us.

_____

_____

You could ask an adult to help with the writing.

12A Pictograms, lists and tables

Explore   Student Book 1, page 161

Ask ten people how they travel to school.

Write the results in the table.

Ways of travelling to school

Way of travelling	Number of people
walk	
bike / scooter	
car	
bus / train	

Use the information in the table to draw a pictogram.

Ways of travelling to school

Way of travelling	Number of people									
walk										
bike / scooter										
car										
bus / train										

Key: ⊗ = I person

Stretch zone

Write an easy question about the information in your pictogram.

_____

Write a hard question about the information in your pictogram.

_____

12B Block diagrams

Explore 2   Student Book 1, page 164

Ask ten people which of these sports they like best.

Write the results in the table.

Favourite sports

Sport	Number of people
football	
cricket	
tennis	
basketball	

Use the information in the table to draw a block diagram.

Favourite sports

Number of people

10
9
8
7
6
5
4
3
2
1

football   cricket   tennis   basketball

Sport

Stretch zone

Write two things that your block diagram shows.

_____

_____

You could ask an adult to help with the writing.

12B Block diagrams

Explore 3   Student Book 1, page 165

Think of four animals.
Write them in the table.

Ask ten people which
animal they like best.

Write the results in the table.

Use the information in the table to
draw a block diagram.

Don't forget to write the labels!

Favourite animals

Animal	Number of people

Favourite animals

Number of
people

10
9
8
7
6
5
4
3
2
1

Animal

Stretch zone

Write an easy question about the information in your block diagram.

_____

Write a hard question about the information in your block diagram.

_____

12 Statistics

147

Discover   Student Book 1, page 166

Draw six different shapes in this diagram. There must be at least one shape in each oval.

Look at the label for each oval before you begin to draw.

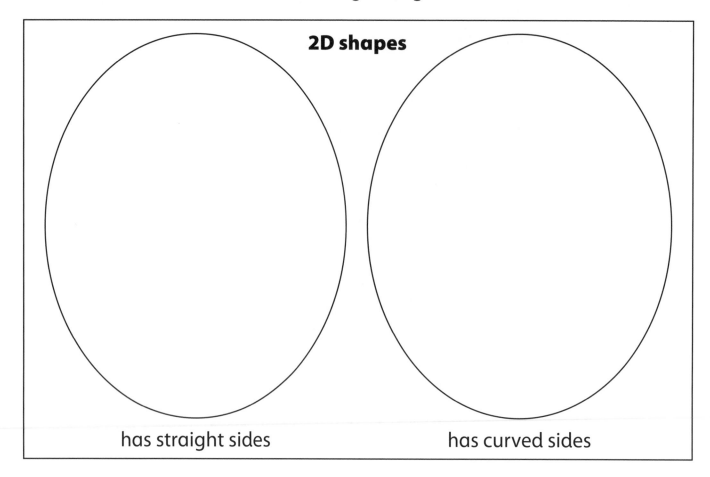

2D shapes

has straight sides   has curved sides

Stretch zone

Choose one shape. Colour it red.

colouring pencil: red

How did you decide where to put that shape?

_____

_____

You could ask an adult to help with the writing.

12C Venn diagrams

Explore   Student Book 1, page 168

Draw arrows to show where each vehicle goes in the Venn diagram.

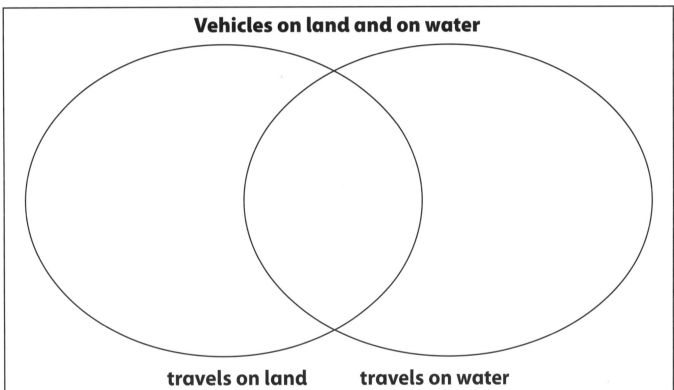

Vehicles on land and on water

travels on land   **travels on water**

 **Stretch zone**

Where did you put the hovercraft? Explain why.

_____

Can you think of a vehicle that you cannot put in this diagram?

_____

Discover   Student Book 1, page 171

Use this Carroll diagram to sort the shapes.

Colour the shapes to show where they go in the Carroll diagram.

> • colouring pencils: red, blue, yellow, green

2D shapes

	Fewer than five sides	**Not fewer than five sides**
All sides straight	Colour these shapes **red**.	Colour these shapes **blue**.
Not all sides straight	Colour these shapes **yellow**.	Colour these shapes **green**.

Stretch zone

Write two things that this Carroll diagram shows.

_____

_____

12D Carroll diagrams

Explore   Student Book 1, page 172

Write the names of eight people you know.

_____   _____   _____   _____

_____   _____   _____   _____

Sort them using a Carroll diagram. You can choose how to sort them.

Don't forget to write the sorting criteria!

		Not
Not		

Stretch zone

Write two differences between a Venn diagram and a Carroll diagram.

_____

_____

Review

1 Draw a face next to each bubble to show how you feel about your learning.

collecting data

reading and creating pictograms

reading and creating block diagrams

Venn diagrams and Carroll diagrams

 2 Tell a partner about one thing you did really well in this unit.

3 Draw or write about things you found easy, challenging or really hard.

What work did you feel confident doing?

What work was challenging?

Is there any work you might need some extra help with?